London Mathematical Society Lecture Note Series. 11

New Developments in Topology

The Edited and Revised
Proceedings of the Symposium on Algebraic Topology,
Oxford, June 1972

J. F. Adams, E. H. Brown Jr. and M. Comenetz, A. Dold,
L. Hodgkin, J. R. Hubbuck, I. Madsen and R. J. Milgram,
J. P. May, D. Quillen, C. T. C. Wall, A. Zabrodsky

Edited by Graeme Segal

Cambridge · At the University Press · 1974

CAMBRIDGE UNIVERSITY PRESS
Cambridge, New York, Melbourne, Madrid, Cape Town, Singapore, São Paulo

Cambridge University Press
The Edinburgh Building, Cambridge CB2 8RU, UK

Published in the United States of America by Cambridge University Press, New York

www.cambridge.org
Information on this title: www.cambridge.org/9780521203548

First published 1974
Re-issued in this digitally printed version 2008

A catalogue record for this publication is available from the British Library

Library of Congress Catalogue Card Number: 73-84323

ISBN 978-0-521-20354-8 paperback

Contents

Preface

In June 1972 a Symposium in Algebraic Topology was held in Oxford. Of the fourteen invited speakers eleven have submitted manuscripts to be published in these Proceedings. In some cases the manuscripts do not correspond closely to the talks that were given, but my object has been to produce a volume of mathematical rather than historical interest. The main theme of the conference was the relationship between generalized cohomology theories, infinite loop spaces, and the symmetric groups, and I hope that we have produced an interesting account of what is being done in that field from diverse points of view.

On behalf of the Organizing Committee I should like to thank all the participants, and not only the speakers, for making the Symposium so successful.

Graeme Segal

OPERATIONS OF THE N^{TH} KIND IN K-THEORY, AND WHAT WE DON'T KNOW ABOUT RP^{∞}

J. F. ADAMS

I. Operations of the nth kind in K-theory

In the old days, if you wanted to solve some concrete problem in homotopy theory, you began by calculating the ordinaty cohomology groups of all the spaces involved. Then you used primary cohomology operations, such as cup-products and the Steenrod operations. If, or when, those didn't yield enough information you tried secondary ones, and then tertiary and higher ones. Of course, if the problem needed tertiary operations you didn't publish the argument in that form because it was too nasty. However, it was sometimes possible to avoid some of the nastiness by using suitable formal machinery like the Adams spectral sequence.

A little later we realised, with great pleasure, that sometimes by using a generalised cohomology theory - perhaps with primary operations - you could successfully tackle a geometrical problem which if done by ordinary cohomology would have needed operations of arbitrarily high kind. It was always conceded that the choice of the cohomology theory most useful for a particular problem might take hard work, or luck, or both. But there was a sort of democratic movement, which proclaimed that every generalised cohomology theory deserved equal rights. For example, Atiyah showed that it is technically possible to teach K-theory before ordinary cohomology. Of course all normal people still did their calculations in ordinary cohomology first, but they were made to feel that they were mere slaves of habit. The philosophy prevailed that all the apparatus of calculation, with which we are so familiar in the ordinary case, should be set up for generalised cohomology. It was conceded that some results like the Kunneth theorem might require restrictive hypotheses. On the other hand, basic things like cohomology operations seem to arise from mere category-theory, and one imagined that they would work in fair generality.

It is now time to explain that the work to be presented is joint work of David Baird and myself. David Baird had been studying classical complex K-theory localised at an odd prime p; and by any standards this is a very good cohomology theory. David arrived at a belief which I will state in the the following simplified form: in this theory you may be able to set up the apparatus of stable tertiary and higher operations, but the information yielded will be exactly zero. It is only fair to say that at first I found this suggestion both implausible and unwelcome. However, I have convinced myself that it is well founded.

To make the suggestion precise, we need machinery to estimate the scope and field of action for stable tertiary and higher operations in a given theory. In the classical case one uses homological algebra, introducing Tor and Ext over the Steenrod algebra. It is natural to try and carry this approach over to the generalised case. Given a spectrum E, you can form $E^*(E)$, the E-cohomology of the spectrum E - this is the graded algebra of stable primary operations on E-cohomology. It has to be considered as a topological algebra. Novikov took this line with success in the case of complex cobordism $E = MU$.

In the case of K-theory I think that Graeme Segal made some tentative calculations with rings of operations generated by operations Ψ^k with k prime to p. But here one suffers from a certain lack of confidence that the algebra is well related to any geometry one can do, so that one is not certain one is doing the right calculations. I believe that when the calculations seemed difficult to interpret, they were abandoned.

Later I suggested that for suitable spectra E one should consider $E_*(E)$, the E-homology of X. Under reasonable assumptions on E this behaves like the dual of the Steenrod algebra, and for any X one can make the E-homology groups $E_*(X)$ into a comodule over the coalgebra $E_*(E)$. One can form Ext of comodules over this coalgebra, and one can be sure that this homological algebra is well related to suitable geometry. (See my lectures in Springer Lecture Notes No. 99).

We are now close to a theorem. Let K be classical complex K-theory localised at an odd prime p. Let X and Y be (say) finite CW-complexes. Then we can form the reduced groups $\tilde{K}_*(X)$, $\tilde{K}_*(Y)$

and regard them as comodules over the coalgebra $K_*(K)$. Sufficient information on $K_*(K)$ has been published by Adams, Harris and Switzer. We can form $\mathrm{Ext}^{s,t}_{K_*(K)}\left(\tilde{K}_*(X),\ \tilde{K}_*(Y)\right)$.

Theorem. $\mathrm{Ext}^{s,t}_{K_*(K)}\left(\tilde{K}_*(X),\ \tilde{K}_*(Y)\right) = 0$ for $s \geq 3$.

This theorem has something to displease everybody. On the one hand, some of us are to some extent true English problem-solvers, and so we like tertiary operations, and when we are told that in this context they are useless, that is a source of pain and grief. On the other hand, some of us are to some extent true French Bourbakistes, and so we hate these inelegant higher operations and would love a theorem which dispenses us from ever considering the nasty things again. But this theorem won't do that, because it says you may still need secondary operations, and we can give examples where you do need them.

Perhaps it will be best to suggest how we should go on from here. The groups $\mathrm{Ext}^{s,t}_{K_*(K)}\left(\tilde{K}_*(X),\ \tilde{K}_*(Y)\right)$ should be the E_2 term of an Adams spectral sequence, but it should not converge to ordinary stable homotopy theory. However, there is a plausible candidate for the groups to which it should converge. To construct them, we start from some category C in which we can do stable homotopy theory. We then construct a category of fractions F. More precisely, F comes equipped with a functor $T : C \to F$, which has the following two properties.

(i) If $f : X \to Y$ is a morphism in C such that $K_*(f) : K_*(X) \to K_*(Y)$ is iso, then $Tf : TX \to TY$ is invertible in F.

(ii) $T : C \to F$ is universal with respect to property (i).

Heuristically, the category F would give an account of stable homotopy-theory, so far as it can be seen through the spectacles of K-theory. For example, it is plausible that an Adams spectral sequence starting from $\mathrm{Ext}^{s,t}_{K_*(K)}\left(K_*(X),\ K_*(Y)\right)$ should converge to the set of morphisms in F from X to Y.

By constructing F we should get a theory with all the same formal properties as stable homotopy theory, but with different coefficient groups. For example, in ordinary stable homotopy theory the group $C[S^{n+r},\ S^n]$ contains a summand Z_p if $r = 2(p-1)q - 1$ with

q prime to p, $q > 0$. I presume (but it is not proved) that in the category of fractions the group $F[S^{n+r}, S^n]$ would actually be Z_p if $r = 2(p-1)q - 1$ with q prime to p, whether q is positive or negative. The price for gaining a certain amount of periodicity is that you lose the Hurewicz theorem.

I would hope that to calculate the set of morphisms $F[X, Y]$ in this category of fractions would be quite reasonable, although we do not yet have the theorems which would enable us to do it. If so, then the category of fractions F might become useful as a computational tool. It is possible that F would retain just enough of the phenomena in C to have some interest.

One final question: should there exist also a version of unstable homotopy-theory as seen through the eyes of K-theory?

II. What we don't know about RP^∞

Here I would like to advertise an unsolved problem, in the hope of throwing it open to wider participation.

When we consider the iterated suspension homomorphism in unstable homotopy groups of spheres we need information about the homotopy groups of truncated real projective spaces RP^{n+r}/RP^n, and in particular about the stable homotopy groups of these spaces. Now, for r fixed the stable homotopy type of RP^{n+r}/RP^n is a periodic function of n; this allows us to speak of the stable homotopy type of RP^{n+r}/RP^n even when n is negative, interpreting it by periodicity; that is, if n is negative we interpret the stable homotopy type of RP^{n+r}/RP^n as that of RP^{n+r+2^m}/RP^{n+2^m} (but shifted down by 2^m dimensions), where m is appropriately large. This periodicity allows us to speak in certain ways which are pleasantly simple and dramatic. For example, it is more memorable to talk about dimension -1, rather than about a positive dimension congruent to -1 modulo 2^m for m appropriately large.

If we filter the spaces RP^{n+r}/RP^n by their subspaces RP^{n+s}/RP^n, and apply stable homotopy, we obtain a spectral sequence. The E^2 term of this spectral sequence has the following form:

$$E^2_{p,q} = \begin{cases} Z_2 \otimes \pi^S_q(S^0) & (p \text{ odd}) \\ \mathrm{Tor}^Z_1(Z_2, \pi^S_q(S^0)) & (p \text{ even}). \end{cases}$$

Here $\pi^S_q(S^0)$ is the qth stable homotopy group of the sphere. I emphasise that this equation is intended to be equally valid for $p > 0$ and for $p \leq 0$. For fixed finite q and r, $E^r_{p,q}$ is periodic in p with period 2^m for some m depending on q and r. The groups $E^\infty_{p,q}$ may be defined, but they are not periodic.

There is some evidence for the following conjecture.

Conjecture (after Mahowald). _This spectral sequence converges to_ $\pi^S_*(S^{-1})$.

The sphere S^{-1} of stable dimension -1 appears because we have

$$RP^{-1}/RP^{-n} \simeq S^{-1} \vee (RP^{-2}/RP^{-n}).$$

This equation, of course, has to be interpreted as a statement about the stable homotopy type of RP^{2^m-1}/RP^{2^m-n} for m sufficiently large, as explained above.

It is tempting to talk about this conjecture in picturesque language, and speak as if there were a spectrum which is like the suspension spectrum of RP^∞, but has one cell in each dimension p whether p is positive, negative or zero. Just as the cohomology ring $H^*(RP^\infty; Z_2)$ is the polynomial ring $Z_2[x]$, where $x \in H^1(RP^\infty; Z_2)$, so the cohomology of this hypothetical spectrum should be the ring of finite Laurent series $Z_2[x, x^{-1}]$. It is now obvious what the cohomology of this hypothetical spectrum should be as a module over the Steenrod algebra. In fact, for $n \geq 0$ we have

$$Sq^i x^n = c(n,i)\, x^{n+i},$$

where the coefficient $c(n, i) \in Z_2$ is a periodic fraction of n; we define the coefficient $c(n, i)$ for negative values of n by periodicity and define $Sq^i x^n$ accordingly. This construction does make $Z_2[x, x^{-1}]$ into a module over the mod 2 Steenrod algebra; in fact, the Adem relations are satisfied, because it is sufficient to observe that they hold on x^n when n is positive. One can now make simple calculations. For example, let $Sq = \sum_{i \geq 0} Sq^i$; then $Sq\, x = x(1 + x)$ and Sq is multiplicative, so $Sq\, x^{-1} = x^{-1}(1 + x)^{-1} = x^{-1} + 1 + x + x^2 \ldots$, that is $Sq^i x^{-1} = x^{i-1}$.

It is necessary to point out firmly that there is no spectrum, in Boardman's category or in any other sensible category, whose cohomology is this A-module. In fact, if there were, then the generator for the 0-dimensional homology group would come from a finite subspectrum, say X. Then we could choose i so large that x^{-i} would restrict to zero in $H^{-i}(X; Z_2)$. But then $(\chi\, Sq^i)x^{-i} = 1$ would restrict to zero in $H^0(X; Z_2)$, a contradiction.

The 'hypothetical spectrum' is therefore a mythical beast. The statements we want to make do not refer to it: they are all to be interpreted in terms of finite complexes RP^{n+r}/RP^n and limits.

One line of thought which might lead one towards the conjecture stated is the consideration of

$$\mathrm{Ext}_A^{**}(Z_2[x, x^{-1}], Z_2).$$

First we consider Ext_{A_n}, where A_n is the subalgebra of A generated by $Sq^1, Sq^2, \ldots, Sq^{2^n}$. As a module over A_n, $Z_2[x, x^{-1}]$ is generated by the powers x^i with $i \equiv -1 \bmod 2^{n+1}$. We may filter $Z_2[x, x^{-1}]$ by the A_n-submodule generated by these generators x^i with $i = 2^{n+1}r - 1$, $r \leq p$. The successive subquotients have the form $A_n \otimes_{A_{n-1}} Z_2$, and their Ext groups are given by

$$\mathrm{Ext}_{A_n}^{**}(A_n \otimes_{A_{n-1}} Z_2, Z_2) \cong \mathrm{Ext}_{A_{n-1}}^{**}(Z_2, Z_2).$$

If we have to conjecture the structure of

$$\mathrm{Ext}_{A_n}^{**}(Z_2[x, x^{-1}], Z_2),$$

the simplest conjecture is

$$\sum_{r \in Z} \mathrm{Ext}^{**}_{A_{n-1}} (Z_2, Z_2),$$

on generators of dimension $2^{n+1}r - 1$. This conjecture is true for $n = 2, 3$. We then have

$$\mathrm{Ext}^{**}_A (Z_2[x, x^{-1}], Z_2) = \lim_{\overleftarrow{n}} \mathrm{Ext}^{**}_{A_n} (Z_2[x, x^{-1}], Z_2);$$

if we make the simplest conjecture for the maps of the inverse system, the inverse limit would be $\mathrm{Ext}^{**}_A (Z_2, Z_2)$, on one generator of dimension -1.

Alternatively, we might proceed as follows. Let M be the sub-A-module of $Z_2[x, x^{-1}]$ generated by the x^i with $i \neq -1$. If M were connected, we would count its generators by computing $Z_2 \otimes_A M = 0$. That is, M is a module 'with no generators'. Similarly, $\mathrm{Tor}^A_1(Z_2, M)=0$; that is, M has 'no relations' between its 'no generators'. If M were connected it would follow that $\mathrm{Ext}^{**}_A (M, Z_2) = 0$; as M is not connected this does not follow, but we might still conjecture it.

Unfortunately, such statements are unlikely to lead anywhere, because even if there is an Adams spectral sequence starting from $\mathrm{Ext}^{**}_A(Z_2[x, x^{-1}], Z_2)$, it is likely to be very hard to prove anything useful about its convergence. We therefore turn to other evidence. In what follows, the groups $E^{\infty}_{p, q}$ are those of the spectral sequence in the conjecture.

Proposition 1. _For_ $q = 0$, _the groups_ $E^{\infty}_{p, 0}$ _are zero except for_ $p = -1$; $E^{\infty}_{-1, 0} = Z_2$.

We can even state the differentials which lead to this state of affairs. Let us write i_n for the homology generator in dimension n (with coefficients Z or Z_2 as may be needed). Then we have

$$\begin{aligned}
d_2 i_p &= \eta i_{p-2} && \text{for } p \equiv 1 \mod 4 \\
d_4 i_p &= \nu i_{p-4} && \text{for } p \equiv 3 \mod 8 \\
d_8 i_p &= \sigma i_{p-8} && \text{for } p \equiv 7 \mod 16 \\
d_9 i_p &= \sigma\eta i_{p-9} && \text{for } p \equiv 15 \mod 32
\end{aligned}$$

$$d_{10} i_p = \sigma\eta^2 i_{p-10} \quad \text{for } p \equiv 31 \mod 64$$

$$d_{12} i_p = \zeta i_{p-12} \quad \text{for } p \equiv 63 \mod 128$$

$$d_{16} i_p = \rho i_{p-16} \quad \text{for } p \equiv 127 \mod 256 \text{ etc.}$$

Proposition 2. *For $q = 1$, the groups $E^{\infty}_{p,1}$ are zero except for $p = -2$; $E^{\infty}_{-2,1} = Z_2$.*

The group for $p = -2$ arises as follows. The complex RP^0/RP^{-3} has the form $(S^{-2} \vee S^{-1}) \cup e^0$, where the components of the attaching map for e^0 are η and 2. So in RP^0/RP^{-n}, the element $2 \in \pi^S_{-1}(S^{-1})$ compresses to RP^{-2}/RP^{-n}, and gives a generator for the group $E^{\infty}_{-2,1}$.

For some distance we can inspect the differentials which lead to this state of affairs. For $p \equiv 3 \bmod 4$, $E^2_{p,1}$ consists of boundaries. We have

$$d_2 \eta i_p = \eta^2 i_{p-2} \quad \text{for } p \equiv 0, 1 \mod 4$$

$$d_6 \eta i_p = \nu^2 i_{p-6} \quad \text{for } p \equiv 2 \mod 8.$$

At this point one guesses that $d_{14} \eta i_p = \sigma^2 i_{p-14}$ for $p \equiv 6 \bmod 16$, but this is wrong. We have

$$d_8 \eta i_p = \bar{\nu} i_{p-8} \quad \text{for } p \equiv 6 \mod 16$$

$$d_9 \eta i_p = \sigma\eta^2 i_{p-9} \quad \text{for } p \equiv 14 \mod 32.$$

It may be as well to point out the first place where we get an element of $\pi_{i-1}(S^{-1})$ with $i > 0$.

Proposition 3. *For $q = 3$ and p odd, the groups $E^{\infty}_{p,3}$ are zero except for $p = -3$; $E^{\infty}_{-3,3} = Z_2$; the element $\eta \in \pi_0(S^{-1})$ compresses to RP^{-3}/RP^{-n} and gives a generator for $E^{\infty}_{-3,3}$.*

We would be on firmer ground, of course, if we could prove that $E^{\infty}_{p,q} = 0$ for $p + q < -1$ - for the conjecture implies this. Applying S-duality to the finite complexes and then passing to the limit, we would like to prove something like this:

Conjecture. *The group of maps in Boardman's category from the suspension spectrum of* RP^∞ *to the suspension spectrum of* S^n *is zero if* $n > 0$.

This would seem to be a stable analogue of one of the following two related conjectures, which are due to Sullivan. For the first, let X be a finite simplicial complex on which Z_2 acts simplicially; let Z_2 act on S^∞ by the antipodal map.

Conjecture. *The evident map, from the fixed-point set of* Z_2 *in* X, *to the function space of equivariant maps from* S^∞ *to* X, *induces an isomorphism of* mod 2 *cohomology.*

Conjecture. *If* Y *is a finite complex, the function-space of base-point-preserving maps from* RP^∞ *to* Y *is contractible.*

The conjectures as stated seem to be inaccessible. Even if we replace them by suitable stable analogues, there is reason to think that no reformulation will be simultaneously convenient to prove and useful for the present purpose.

Finally, the Kahn-Priddy theorem is relevant to the present conjecture; it proves that for $p = -1$, $E^\infty_{-1,q}$ is zero for $q > 0$. It is just possible that comparable methods, based on the consideration of infinite loop-spaces, would be helpful in studying the present conjecture. The plan would be to take the suspension-spectrum $(RP^N/RP^{-n})/S^{-1}$, replace it by an Ω-spectrum, obtain information on the homology of the loop-spaces in the Ω-spectrum, and pass to limits.

Department of Pure Mathematics and Mathematical Statistics
16 Mill Lane
Cambridge CB2 1SB

THE PONTRJAGIN DUAL OF A SPECTRUM

EDGAR H. BROWN, JR. and MICHAEL COMENETZ[†]

Recall, if G is a discrete abelian group and $c(G)$ is its character group, that is,

$$c(G) = \mathrm{Hom}(G, R/Z),$$

then Pontrjagin duality provides an isomorphism

$$H^q(X; c(G)) \approx c(H_q(X; G)).$$

The aim of this talk is to describe, without proofs, how this duality can be incorporated into generalized homology and cohomology theories and spectra.

Regarding spectra, we work in Boardman's graded homotopy category of spectra $\mathcal{S}_{h*}$ ([5]). We denote the morphisms of degree q $(f: A \to S^q B)$ by

$$\{A, B\}_q.$$

If A is a spectrum and X is a CW complex or a spectrum, $A_q(X)$ and $A^q(X)$ denote the homology and cohomology of X with coefficients in A, respectively.

Let $\mathcal{G}$ be the category of discrete abelian groups. We ignore the topology on $c(G)$ so that $c: \mathcal{G} \to \mathcal{G}$. Recall c takes exact sequences into exact sequences and direct sums into direct products. Hence for any spectrum A, $c(A_*(\ ))$ is an additive cohomology theory on the category of all CW complexes. Therefore by [2], there is a spectrum A' and a natural equivalence

---

† While doing this work E. H. Brown was supported by NSF Grant GP 28938. Some of the material presented here is part of M. Comenetz s Ph. D Thesis.

$$t_A: (A')^*(\ ) \approx c(A_*(\ )).$$

The main results of this talk concern the relation between A and A'. We do not know how to adequately deal with the natural topology on $c(A_*(\ ))$. In consequence, most of our results on A' require that we assume $\pi_i(A)$ is finite for all i.

Let S be the sphere spectrum. Choose a spectrum $c(S)$ and a natural equivalence

$$t_S: c(S)^*(\ ) \approx c(S_*(\ )).$$

We identify $c(S)^0(S^0)$ with R/Z via the isomorphism:

$$c(S)^0(S^0) \overset{t_S}{\approx} c(S_0(S^0)) \approx c(Z) = R/Z.$$

Definition (1. 1). A <u>duality</u> between spectra A and A' is a pairing

$$\mu: A \wedge A' \to c(S)$$

such that the Kronecker index

$$A_q(X) \otimes (A')^q(X) \to c(S)^0(S^0) = R/Z$$

induces an isomorphism

$$t_\mu: (A')^q(X) \approx c(A_q(X))$$

for all CW complexes X and all q. We call A' a dual of A.

The existence and uniqueness of duals and their functorial properties are contained in the following:

Theorem (1. 2). A' <u>is a dual of</u> A <u>if and only if</u> A' <u>represents the function spectrum</u> $F(A, c(S))$ (see [5] for the definition of F).

Definition (1. 3). If $\mu_i: A_i \wedge A_i' \to c(S)$, $i = 1, 2$, are dualities and $f: A_1 \to A_2$ is a map of degree q, the <u>dual</u> of f with respect to μ_1 and μ_2 is a map $f': A_2' \to A_1'$ of degree q such that

$$\mu_2(f \wedge id_{A_2}) = \mu_1(id_{A_1} \wedge f').$$

The existence and uniqueness of f' follows from (1. 2).

We make the process of passing from A to A' into a functor as follows. For each spectrum A choose a spectrum $c(A)$ and a duality

$$\mu_A: A \wedge c(A) \to c(S).$$

If $f: A \to B$, let $c(f): c(B) \to c(A)$ denote the dual of f with respect to μ_A and μ_B. The following is immediate:

Theorem (1. 4).

$$c: \mathcal{S}_{h_*} \to \mathcal{S}_{h_*}$$

is a contravariant functor which preserves the degree of morphisms.

In general $c(c(A))$ and A are not isomorphic because we have ignored the natural topology on $c(A_*(\ ))$. Note if X is a finite CW complex and $\pi_i(A)$ is finite for all i, $A_q(X)$ is finite and hence $c(c(A_q(X))) \approx A_q(X)$. This enables us to prove that $c(c(A)) \approx A$ if $\pi_i(A)$ is finite.

Let $\mathcal{S}_f$ be the full subcategory of $\mathcal{S}_{h_*}$ of all spectra A such that $\pi_i(A)$ is finite for all i. Note $c: \mathcal{S}_f \to \mathcal{S}_f$. We next list some properties of c on $\mathcal{S}_f$.

Theorem. In the following all spectra are in $\mathcal{S}_f$.

(1. 5) $cc: \mathcal{S}_f \to \mathcal{S}_f$ is naturally equivalent to the identity functor.

(1. 6) $c: \{A, B\}_q \approx \{c(B), c(A)\}_q$.

(1. 7) $\pi_i(c(A)) \approx c(\pi_{-i}(A)) \approx \pi_{-i}(A)$.

(1. 8) If $H(G)$ is the Eilenberg-MacLane spectrum with

$$\pi_i(H(G)) = \begin{cases} G & i = 0 \\ 0 & i \neq 0 \end{cases}$$

then

$$c(H(G)) = H(c(G)).$$

(1.9) _If p is a prime, choose_ $c(H(Z_p)) = H(Z_p)$. _Note_

$$\mathcal{A}_p = \{H(Z_p), H(Z_p)\}_*$$

is the Steenrod algebra. Then

$$c: \mathcal{A}_p \to \mathcal{A}_p$$

is the usual canonical antiautomorphism of the Steenrod algebra.

(1.10) _If_

$$A \xrightarrow{f} B \xrightarrow{g} C \xrightarrow{h} A$$

is an exact triangle in the sense of [5],

$$c(A) \xrightarrow{c(h)} c(C) \xrightarrow{c(g)} c(B) \xrightarrow{-c(f)} c(A)$$

is an exact triangle.

We next describe how the ordinary Z_p cohomology of A and c(A) are related when A is connected. As above, let $\mathcal{A}_p$ be the Steenrod algebra. Let c_p be the contravariant functor on the category of graded left $\mathcal{A}_p$ modules to itself given by

$$c_p(M) = \operatorname{Hom}_{\mathcal{A}_p}(M, \mathcal{A}_p).$$

$\operatorname{Hom}_{\mathcal{A}_p}$ is the graded hom; maps of degree q raise dimensions by q. The left $\mathcal{A}_p$ module structure on $c_p(M)$ is given by

$$(au)(m) = u(m)c(a).$$

If A is a spectrum let

$$\lambda: \{H(Z_p), A\}_* \to c_p(H^*(A; Z_p))$$

be given by

$$\lambda(f) = H^*(f; Z_p): H^*(A; Z_p) \to H^*(H(Z_p); Z_p) = \mathcal{A}_p.$$

Note if $A \in \mathcal{S}_f$, c gives an isomorphism

$$H^*(c(A); Z_p) = \{c(A), H(Z_p)\}_* \overset{c}{\approx} \{H(Z_p), A\}_*.$$

Theorem (1.11). _If_ $A \in \mathcal{S}_f$ _and_ $\pi_i(A) = 0$ _for_ $i < N$, _then_

$$\lambda c: H^*(c(A); Z_p) \approx c_p(H^*(A); Z_p)$$

_is an_ $\mathfrak{A}_p$ _linear isomorphism._

Corollary (1.12). _If_ $A \in \mathcal{S}_f$, $\pi_i(A) = 0$ _for_ $i < N$ _and_ $H^q(A; Z) = 0$ _for_ $q > M$, _(for example,_ $A = \{X, SX, S^2X, \ldots\}$ _where_ X _is a finite_ CW _complex with finite homology) then_

$$H^*(c(A); Z) = 0.$$

We conclude this talk with an example of how c may be utilized to define the dual of a stable, higher order cohomology operation. A detailed treatment of duality and higher order operations will be given by the second author in another paper.

Let $\mathcal{C}$ be a free chain complex over $\mathfrak{A}_p$ of the form

$$C_2 \xrightarrow{d_2} C_1 \xrightarrow{d_1} C_0,$$

where C_0 and C_2 are each generated by one element x_0 and x_2. Suppose C_1 is generated by x_{1j}, $i = 1, 2, \ldots, k$ and

$$d_1 x_{1j} = b_j x_0,$$

$$d_2 x_2 = \Sigma a_j x_{1j},$$

a_j, $b_j \in \mathfrak{A}_p$. Suppose $|x_0| = 0$, $|d_1| = |d_2| = 1$ ($|\ \ | =$ degree). In other words $\mathcal{C}$ is the chain complex associated with the relation

$$\Sigma a_j b_j = 0.$$

Let K_i be Eilenberg-MacLane spectra such that $\pi_*(K_i)$ is in Z_p module and

$$H^*(K_i; Z_p) \approx C_i.$$

Let $\alpha_1: K_0 \to K_1$ and $\alpha_2: K_1 \to K_2$ realize d_1 and d_2.

We may form a diagram of spectra

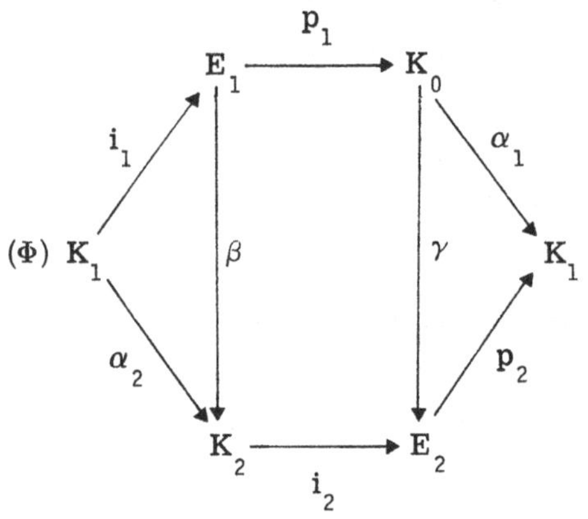

where the top and bottom rows are exact triangles and the diagram commutes. Choose $|i_1| = |p_1| = |p_2| = |i_2| = 0$. E_1 is the fibration over K_0 with fibre K_1 and k-invariant α_1 and similarly for E_2. Φ gives a secondary cohomology operation as follows: (If $F : A \to B$, $F_* : A_q(X) \to B_q(X)$.)

$$\Phi\colon K_0^q(X) \cap \ker(\alpha_1)_* \to K_2^{q+1}(X)/\operatorname{image}(\alpha_2)_*$$

i.e., $\quad \Phi\colon H^q(X) \cap (\cap \ker b_j) \to H^{q+r}(X)/\Sigma\, a_j H^*(X)$,

where

$$r = |x_2| + 1.$$

Φ is given by

$$\Phi(u) = \beta_*((p_{1*})^{-1}u) = (i_2)_*^{-1}\gamma_* u.$$

Applying c_p to the complex $\mathfrak{C}$ we obtain a complex $c_p(\mathfrak{C})$:

$$c_p(C_0) \xrightarrow{c_p(d_1)} c_p(C_1) \xrightarrow{c_p(d_2)} c_p(C_2).$$

One easily checks that $c_p(C_1)$ is a free module on generators $\bar{x}_{1_j}$ where

$$\bar{x}_{1_j}(x_{1i}) = 1, \quad i = j,$$
$$= 0, \quad i \neq j.$$

Similarly, $c_p(C_0)$ and $c_p(C_2)$ are free on generators $\bar{x}_0$ and $\bar{x}_1$ and

$$c_p(d_2)\bar{x}_{1_j} = c(a_j)\bar{x}_2,$$

$$c_p(d_1)\bar{x}_0 = \Sigma\, c(b_j)\bar{x}_{1_j}.$$

Hence $c_p(\mathcal{C})$ is the chain complex associated to the relation $\Sigma\, c(b_j)c(a_i) = 0$.

Applying c to the diagram Φ we obtain a diagram $c(\Phi)$

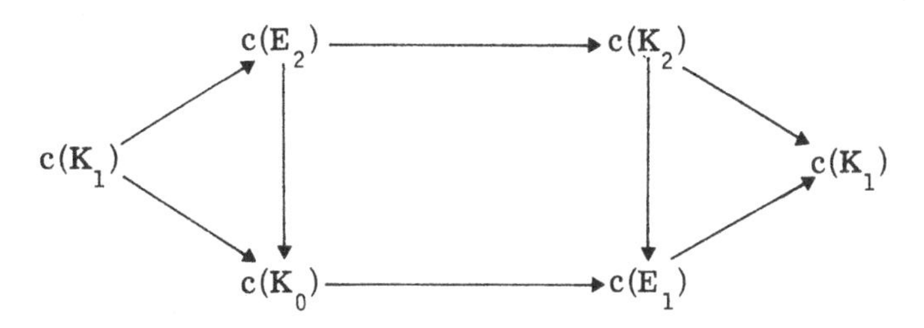

$c(K_i)$ are Eilenberg-MacLane spectra, $H^*(c(K_i); Z_p) \approx c_p(C_i)$, $c(E_2)$ is a fibration over $c(K_2)$ with fibre $c(K_1)$ and k-invariant $c(\alpha_2)$ and the diagram gives the secondary operation associated to the relation $\Sigma\, c(b_j)c(a_i) = 0$.

As cohomology operations, Φ and $c(\Phi)$ are related to one another as follows: Let X be a finite CW complex and let

$$\nu: X \wedge Y \to S^m$$

be a Spanier Whitehead duality. For any spectrum A, ν and μ_A define a non-singular pairing

$$A^q(X) \otimes c(A)^{m-q} \to R/Z.$$

Denote this pairing by $\langle , \rangle$.

Theorem (1.13). $\langle , \rangle$ defines non-singular pairings between the domain of Φ and the range of $c(\Phi)$ and between the range of Φ and the domain of $c(\Phi)$, and

$$\langle \Phi(u), v\rangle = \langle u, c(\Phi)v\rangle .$$

Cohomology operations Φ and $c(\Phi)$ as in (1.13) have been defined by Peterson and Stein in [4] and by Maunder in [3].

REFERENCES

1. J. F. Adams. On the non-existence of elements of Hopf invariant one. Ann. of Math. (1960), 384.
2. E. H. Brown. Cohomology theories . Ann. of Math. 75 (1962), 467-484.
3. C. R. F. Maunder. Cohomology operations and duality . Proc. Camb. Phil. Soc. 64 (1968), 15-30.
4. F. S. Peterson and N. Stein. The dual of a secondary cohomology operation . Ill. J. Math. (1960), 602.
5. R. Vogt. Boardman's stable homotopy category, Lecture Notes Series No. 21, Matematisk Institut, Aarhus Universitet (1970).

Brandeis University
Waltham, Massachusetts 02154
U.S.A.

ALGEBRAIC K-THEORY OF NON-ADDITIVE FUNCTORS OF FINITE DEGREE

ALBRECHT DOLD

This is mainly a summary and commentary of the article [3] with the same title. Proofs are generally omitted here, and so is the treatment of relative K-theory. On the other hand, section (F) gives an application to K_1 which is not contained in [3].

(A) If R is a ring with unit let $\mathcal{P}R$ denote the category of finitely generated projective left R-modules. If, furthermore, I is a small category let $(I; \mathcal{P}R)$ denote the category of covariant functors $Q: I \to \mathcal{P}R$. Thus, $(I; \mathcal{P}R) = \mathcal{P}R$ if I is the category with precisely one morphism; if I has precisely one object (but several morphisms) then the objects of $(I; \mathcal{P}R)$ are finitely generated projective R-modules with operators in $\mathrm{Mor}(I)$.

Recall [7; II. 1] that $K(I; \mathcal{P}R)$ is the abelian group with generators $[Q]$ where $Q \in \mathrm{Ob}(I; \mathcal{P}R)$, and relations $[Q] = [Q'] + [Q'']$ whenever there is an exact sequence $0 \to Q' \to Q \to Q'' \to 0$. Exactness means that $0 \to Q'(i) \to Q(i) \to Q''(i) \to 0$ is (split-) exact for every $i \in \mathrm{Ob}(I)$.

If $h: R \to S$ is a homomorphism of rings then the induced functor $(I; \mathcal{P}h):(I; \mathcal{P}R) \to (I; \mathcal{P}S)$ preserves exact sequences (is exact) and therefore induces a homomorphism $K(I; \mathcal{P}h) : K(I; \mathcal{P}R) \to K(I; \mathcal{P}S)$. More generally, every additive functor $f : \mathcal{P}R \to \mathcal{P}S$ induces a homomorphism $K(I; f) : K(I; \mathcal{P}R) \to K(I; \mathcal{P}S)$. One of the main purposes of the lecture is to generalize this functoriality to non-additive functors $F : \mathcal{P}R \to \mathcal{P}S$ of <u>finite degree</u>. The main tool is a description of the <u>set</u> $K(I; \mathcal{P}R)$ (in terms of semi-simplicial objects) which does not use sums of morphisms or objects.

(B) Let Y denote a semi-simplicial (=s. s) object in $(I; \mathcal{P}R)$. Equivalently, Y is a functor from I to the category of s. s. objects in $\mathcal{P}R$.

Let $NY = Y/\{\text{degeneracies}\}$, the normalized complex of Y. In every dimension, N_jY is a direct summand of Y_j, hence an object of $(I; \mathcal{P}R)$, and we can take its class $[N_jY] \in K(I; \mathcal{P}R)$. If for some natural number d we have $N_jY = 0$ for $j > d$ then we say, $(\dim Y) \leq d$. In this case we can form <u>the characteristic</u>

$$\chi Y = \sum_{j=0}^{\infty} (-1)^j [N_jY]$$

in $K(I; \mathcal{P}R)$. It has the following properties.

(i) $Y \sim Y' \Rightarrow \chi Y = \chi Y'$,

(ii) $Y \simeq Y' \Rightarrow \chi Y = \chi Y'$,

where $\simeq$ denotes homotopy equivalence, and $\sim$ indicates that there are isomorphisms $Y_j \cong Y'_j$ which commute with degeneracy operators (but not necessarily with face operators).

Property (i) is obvious, and (ii) has an easy proof (via the mapping cone). In fact, both properties hold in much more general categories for which one has a K-group. The following converse, however, uses the structure of $(I; \mathcal{P}R)$ more essentially.

Proposition. <u>If</u> $\chi Y = \chi Y'$ <u>then</u> Y, Y' <u>are related by a finite number of steps</u> $\sim$ <u>or</u> $\simeq$.

(C) **Comments.** One easily sees that every $y \in K(I; \mathcal{P}R)$ is of the form $y = \chi Y$ for some s. s. object Y in $(I; \mathcal{P}R)$, with $(\dim Y) \leq 1$. It follows that $K(I; \mathcal{P}R)$ can be obtained by taking the set of all finite-dimensional s. s. objects Y in $(I; \mathcal{P}R)$ and factoring this set by the equivalence relation which is generated by $\sim$ and $\simeq$. The same is true if one considers Y with $(\dim Y) \leq d$ only, provided $d \geq 1$. This result is the s. s. analog of Atiyah's complex-description of K-theory (cf. [1], thm. 2.6.1; compare also [6], §4/5), the advantage being that, in s. s. theory, homotopy does not involve addition.

(D) If $f : A \to B$ is a function between abelian groups (for A, abelian monoid is enough) then its <u>deviations</u> are the functions $f_n : A^n \to B$, $n \geq 0$, defined by $f_0 = f(0)$,

$$f(\sum_{i=1}^{n} a_i) = f(0) + \sum_{k=1}^{n} \sum_{i_1<\ldots<i_k} f_k(a_{i_1}, \ldots, a_{i_k}).$$

If $f_{m+1} = 0$ then $f_n = 0$ for all $n > m$; in this case one says (degree $f) \le m$.

Similarly, if $F : \mathcal{A} \to \mathcal{B}$ is a functor between abelian categories ($\mathcal{A}$ additive and $\mathcal{B}$ amenable is enough) then its cross-effects [5; §9] are the functors $F_n : \mathcal{A} \times \ldots \times \mathcal{A} \to \mathcal{B}$, $n \ge 0$, which are essentially defined by $F_0 = F(0)$,

$$F(\bigoplus_{i=1}^{n} A_i) = F(0) \oplus \bigoplus_{k=1}^{n} \bigoplus_{i_1<\ldots<i_k} F_k(A_{i_1}, \ldots, A_{i_k}).$$

If $F_{m+1} = 0$ then $F_n = 0$ for all $n > m$; in this case one says (degree $F) \le m$.

If $F : \mathcal{A} \to \mathcal{B}$ is any functor and Y is an s. s. object in $\mathcal{A}$ then FY is an s. s. object in $\mathcal{B}$ defined by $(FY)_j = F(Y_j)$, and similarly for face and degeneracy operators. One has the following properties.

(i) $Y \sim Y' \Rightarrow FY \sim FY'$,

(ii) $Y \simeq Y' \Rightarrow FY \simeq FY'$,

(iii) $\dim(FY) \le (\text{degree } F) \cdot (\dim Y)$.

Of these, only (iii) is not obvious (cf. [4], 4. 23).

Property (iii) tells us in particular, that functors of finite degree take finite dimensional s. s. objects into finite-dimensional s. s. objects. Together with (B) we now get the following results.

(E) **Theorem 1.** _If_ R, S _are rings and_ $F : \mathcal{P}R \to \mathcal{P}S$ _is a functor of finite degree then for every small category_ I _there is a unique function of finite degree_

$$K(I; F) : K(I; \mathcal{P}R) \to K(I; \mathcal{P}S)$$

such that $K(I; F)[Q] = [FQ]$, _for all_ $Q \in Ob(I; \mathcal{P}R)$. _Moreover,_ (degree $K(I; F)) \le$ (degree F).

Theorem 2. _If_ F _and_ I _are as in Theorem 1 and_ Y _is a finite-dimensional_ s. s. _object in_ $(I; \mathcal{P}R)$ _then_ FY _is a finite-dimensional_ *s. s. object in* $(I; \mathcal{P}S)$, *and*

$$\chi(FY) = (K(I; F))(\chi Y).$$

Addendum. *If* F *and* I *are as in Theorem 1, and* $P, Q \in Ob(I; \mathcal{P}R)$ *then*

$$(K(I; F))([P] - [Q]) = [FP] + \sum_{j=1}^{\infty} (-1)^j [F_j(Q, \ldots, Q) \oplus F_{j+1}(P, Q, \ldots, Q].$$

The addendum is significant because every element in $K(I; \mathcal{P}R)$ is of the form $[P] - [Q]$. Theorem 2 is really the existence proof for Theorem 1, i. e. the definition of $K(I; F)$, and it should be clear from (B) and (D). The uniqueness part of Theorem 1 has nothing to do with functors and K-theory; it is contained in the following elementary

Lemma. *If* $\alpha : A \to B$ *is a map of degree* $\leq m$ *between abelian groups and if* $G \subset A$ *is a set of generators such that* $\alpha(\sum_{i=1}^{m} g_i) = 0$ *for all m-tuples* $(g_1, \ldots, g_m)$ *in* $G \cup \{0\}$ *then* $\alpha = 0$.

(F) There are also *relative* versions of the preceding results. These, and their proofs, are a little more technical, and we therefore refer the reader to [3]. As to *higher* K-groups K_1R, K_2R, ..., it seems to me that functors $F : \mathcal{P}R \to \mathcal{P}S$ of finite degree should induce *homomorphisms* (and not just maps of finite degree) $K_jF : K_jR \to K_jS$. For $j = 1$ we will prove this property now, whereas for $j > 1$ I have not properly understood K_jR yet.

Remark first that $K(I; \mathcal{P}R)$ is *also a* (contravariant) *functor of* I: Every functor $I' \to I$ induces a homomorphism $K(I; \mathcal{P}R) \to K(I'; \mathcal{P}R)$. Moreover, these homomorphisms commute with the maps $K(I; F)$ of theorem 1, i. e. $K(I; \mathcal{P}R)$ is a *bifunctor*. In particular, we consider $K(G^j; \mathcal{P}R)$, where G^j is the free (non-abelian) group on j generators $\gamma_0, \gamma_1, \ldots, \gamma_{j-1}$. We think of G^j as a category with one object; an object in $(G^j; \mathcal{P}R)$ is then a finitely generated projective R-module M together with j automorphisms $\alpha_0, \ldots, \alpha_{j-1}$ of M. The groups (categories) G^j fit into a familiar co-semi-simplicial pattern

$$G^* = \{G^0 \rightleftarrows G^1 \rightleftarrows G^2 \overset{\eta}{\underset{\varepsilon}{\vdots}} G^3 \ldots\},$$

where the homomorphism $\eta_k : G^{n+1} \to G^n$ takes γ_k into 1 and maps the other generators onto generators, in increasing order; the homomorphism $\varepsilon_k : G^n \to G^{n+1}$ for $0 < k \leq n$ takes γ_{k-1} into $\gamma_{k-1}\gamma_k$ and maps the other generators onto the other generators, in increasing order; whereas $\varepsilon_0(\gamma_j) = \gamma_{j+1}$, $\varepsilon_{n+1}(\gamma_j) = \gamma_j$. Applying $K(-; \mathcal{P}R)$ we obtain an s. s. abelian group

$$K(G^*; \mathcal{P}R) = \{K(G^0; \mathcal{P}R) \rightleftarrows K(G^1; \mathcal{P}R) \rightleftarrows K(G^2; \mathcal{P}R) \vdots K(G^3; \mathcal{P}R) \ldots\}.$$

Since K is a bifunctor, every $F : \mathcal{P}R \to \mathcal{P}S$ of finite degree induces an s. s. map

$$K(G^*; F) : K(G^*; \mathcal{P}R) \to K(G^*; \mathcal{P}S).$$

This is not a homomorphism (degree $K(G^*; F) \leq$ degree F) but it induces homomorphisms of homotopy groups $\pi_j K(G^*; \mathcal{P}R)$ for $j > 0$.

The definition of $K_1 R$ [7; pp. 193-194] easily implies $\pi_1 K(G^*; \mathcal{P}R) \cong K_1 R$, hence we obtain an induced homomorphism for K_1, as promised. Slightly more general, we get

Theorem 3. _If_ F _and_ I _are as in Theorem 1 then the assignment_ $[Q, \alpha] \mapsto [FQ, F\alpha]$ _defines a homomorphism_

$$K_1(I; \mathcal{P}R) \to K_1(I; \mathcal{P}S),$$

where K_1 _of the category_ $(I; \mathcal{P}R)$ _is defined as in_ [7; p. 193], _with generators_ $[Q, \alpha]$ _such that_ $Q \in Ob(I; \mathcal{P}R)$, $\alpha \in Aut(Q)$.

Indeed, if I consists of just one morphism then this is what we proved above; in general, we simply replace $\mathcal{P}R$ by $(I; \mathcal{P}R)$ in the above argument.

(G) What about the other homotopy groups of $K(G^*; \mathcal{P}R)$, or of $K(G^*; (I; \mathcal{P}R)) = K(G^* \times I; \mathcal{P}R)$? We have already remarked that the groups $\pi_j K(G^*; \mathcal{P}R)$, for $j > 1$, are also abelian-group-valued functors with respect to morphisms $F : \mathcal{P}R \to \mathcal{P}S$ of finite degree. They look interesting to me and should be related to K_jR, but how? For $j = 0$ we have, of course, $\pi_0 K(G^*; \mathcal{P}R) = KR$, and the induced map (not homomorphism) $\pi_0 K(G^*; F)$ coincides with the map KF of Theorem 1.

(H) For many rings R, S the rank function defines isomorphisms $KR \cong \mathbf{Z} \cong KS$. A functor $F : \mathcal{P}R \to \mathcal{P}S$ of degree $\leq n$ then induces a function $KF : \mathbf{Z} \to \mathbf{Z}$ of degree $\leq n$, hence KF is a linear combination of functions $z \mapsto \binom{z+j}{j}$, where $0 \leq j \leq n$. The coefficients $(KF)_j \in \mathbf{Z}$ of this linear combination can be calculated from the following system of linear equations (which has determinant 1)

$$\sum_{j=0}^{n} (KF)_j \binom{k+j}{j} = \operatorname{rank}(FR^k), \quad 0 \leq k \leq n;$$

i.e. one has to determine the rank of $F(R^k)$ for $k = 0, 1, \ldots, n$, and then solve for the unknowns $(KF)_j$. Algebraically, this may not be very exciting, geometrically however, Theorem 2 does give interesting results for the Euler-characteristic. Consider for instance, ρ-powers of spaces Y, where $\rho : \Gamma \to S(n)$ is a representation of a group Γ in the symmetric group; by definition, the ρ-power $P^\rho Y$ is the quotient of $Y \times Y \ldots \times Y$ (n-factors) under the action of Γ (via ρ). This functor P^ρ of spaces corresponds to the analogous functor $P^\rho A = (\otimes^n A)/\Gamma$ of (free) abelian groups under geometric realization [2; §7], and here degree $(P^\rho) = n$. Thus, theorem 2 has the following

Corollary (to Theorem 2). _If $\rho : \Gamma \to S(n)$ is a representation then there are (unique) integers $\rho_j \in \mathbf{Z}$, $0 \leq j \leq n$, such that_

$$\chi(P^\rho Y) = \sum_{j=0}^{n} \rho_j \binom{\chi Y + j}{j}$$

for all compact CW-spaces Y, where χ = Euler-characteristic.

In order to find ρ_j one can determine $\chi P^\rho(k)$ = number of points in $P^\rho(k)$, where (k) is a finite set with k points, and solve the linear system

$$\sum_{j=0}^{n} \rho_j \binom{k+j}{j} = \chi\, P^{\rho}(k), \qquad 0 \leq k \leq n.$$

(I) A simple example with operators is obtained if R is a field and $I = \mathbf{N}$ = additive monoid of natural numbers (a category with one object). An object in $(I; \mathcal{P}R)$ is then a pair (V, α) where V is a finite-dimensional vector space over R and $\alpha : V \to V$ is a linear map. The characteristic polynomial $\Phi(\alpha; t) = \det(t - \alpha)$ defines an isomorphism

$$\Phi : K(\mathbf{N}; \mathcal{P}R) \cong R\{t\}^*/R^*,$$

where $R\{t\}$ = field of rational functions and * denotes the multiplicative groups. The characteristic χY as used in Theorem 2 is now the characteristic rational function $\Phi(\alpha; t)$, where $Y = (X, \alpha)$, X a finite-dimensional s. s. vector-space, $\alpha : X \to X$ an s. s. linear map, and

$$\Phi(\alpha, t) = \det(t-\alpha_{\text{even}})/\det(t-\alpha_{\text{odd}}) = \det(t-(H\alpha)_{\text{even}})/\det(t-(H\alpha)_{\text{odd}}),$$

where H = homology. As in (H), there are applications to geometry (Lefschetz-numbers, etc.); we refer the reader to [3; §5].

REFERENCES

1. M. Atiyah. K-theory. Benjamin, New York, 1968
2. A. Dold. Homology of symmetric products and other functors of complexes. Ann. of Math. 68 (1958), 54-80.
3. A. Dold. K-theory of non-additive functors. Math. Annalen 196 (1972), 177-197.
4. A. Dold and D. Puppe. Homologie nicht-additiver Funktoren. Ann. Inst. Fourier 11 (1961), 201-312.
5. S. Eilenberg and S. MacLane. On the groups $H(\pi, n)$ II. Ann. of Math. 60 (1954), 49-139.
6. J. L. Kelley and E. H. Spanier. Euler-characteristics. Pac. J. of Math. 26 (1968), 317-339.

7. R. Swan. Algebraic K-theory. Lect. Notes in Math. 76 (1968). Springer, Heidelberg.

Mathematisches Institut
Universität Heidelberg
6900 Heidelberg 1
Im Neuenheimer Feld 9
Germany

DYER-LASHOF OPERATIONS IN K-THEORY

LUKE HODGKIN

In this talk I shall discuss a family of homology operations $\{Q^{(p)}\}$ operating in the K-theory mod p of infinite loopspaces - one for each p - which, in a sense which I hope to make clear, correspond to the whole system of Dyer-Lashof operations [1] in ordinary homology. (p is an odd prime throughout - certain changes need to be made for the mod 2 theory as usual.) It seems difficult to axiomatize homology operations in infinite loopspaces, but the Dyer-Lashof operations are, in some sense, the only ones there are, and they are quite well understood by now (see [3, 4, etc.]). Among things that we can use to characterize them are their naturality properties; their method of construction (dual to Steenrod operations, using the structure maps

$$\mu_p : E\mathfrak{S}_p \times_{\mathfrak{S}_p} X^p \to X$$

of H^∞-spaces); and the fact that suitable words in the Dyer-Lashof operations on $H_*(X; \mathbf{Z}/p)$ generate $H_*(QX; \mathbf{Z}/p)$ as an algebra, where $QX = \varinjlim \Omega^n S^n X$.

I'll begin by pointing out some of the difficulties in applying this to K-theory, in reference to how my own ideas have developed. First, if we consider the important H^∞-space $(QS^0)_0$ (identity component), the results of [2] imply that $K_*((QS^0)_0; \mathbf{Z}/p)$ is a polynomial algebra on a single set of generators $\{\tilde{\theta}_{p^r}\}$ $(r = 1, 2, \ldots)$ where in the notation of [2] $\tilde{\theta}_{p^r} = \theta_{p^r}\theta_1^{-p^r}$. It would seem plausible - from other facts as well - that these should be, modulo decomposables, the iterates of a single operation $Q^{(p)}$ on $\tilde{\theta}_p$, i.e. that we should have

$$\tilde{\theta}_{p^r} = (Q^{(p)})^r(\tilde{\theta}_p) + \text{decomposables}.$$

Unfortunately calculations with the diagonal formula of [2] give the following.

Proposition 1. The $(p + 1)^{st}$ power map of $(QS^0)_0$, which is a map of infinite loopspaces, leaves fixed $\tilde{\theta}_p$ but not any of the other $\tilde{\theta}_{p^r}$'s.

Pursuing further the implications of this result - which it is difficult to pin down as definitely contradicting the existence of Dyer-Lashof operations in itself - I arrived at a second and more serious indication of a problem. It is basic in Dyer and Lashof's method of constructing the operations that the homology of $E\mathfrak{S}_p \times_{\mathfrak{S}_p} X^p$ is a functor of the homology of X. But, writing $K_*^{\pi}(X^p; \mathbf{Z}/p)$ for $K_*(E\pi \times_{\pi} X^p; \mathbf{Z}/p)$ when π is a subgroup of $\mathfrak{S}_p$, we have the following.

Theorem 1. There exists a finite CW complex X and map $f : X \to X$ such that f induces the identity on $K_*(X; \mathbf{Z}/p)$ but not on $K_*^{\mathfrak{S}_p}(X^p; \mathbf{Z}/p)$. Hence, $K_*^{\mathfrak{S}_p}(X^p; \mathbf{Z}/p)$ is not a functor of $K_*(X; \mathbf{Z}/p)$.

I shall outline a proof of this - the counter-example is linked with Proposition 1 - later. It is time, though, to look on the positive side; the situation looks, in terms of what we know of $K_*(QS^0)$ etc., too encouraging for there to be no way out, and we can find one by passing to 'secondary' operations.

Theorem 2. For each $p \neq 2$ there is a natural transformation on the category of H^{∞}-spaces

$$Q^{(p)} : \tilde{K}_*(X; \mathbf{Z}/p) \to \tilde{K}_*(X; \mathbf{Z}/p)/(\tilde{K}_*(X; \mathbf{Z}/p))^p$$

which arises as the composition of μ_{p*} with a natural transformation ϕ_* from $\tilde{K}_*(X; \mathbf{Z}/p)$ to a suitable quotient of $K_*^{\mathfrak{S}_p}(X; \mathbf{Z}/p)$.

Also, in the notation of [2],

$K_*(QS^0; \mathbf{Z}/p) = \mathbf{Z}/p[\theta_1, \theta_1^{-1}, \theta_p, \theta_{p^2}, \ldots]$ with

$$Q^{(p)}(\theta_{p^r}) = \theta_{p^{r+1}} + (\tilde{K}_*(QS^0; \mathbf{Z}/p))^p. \quad (r = 0, 1, 2, \dots)$$

Note. An improvement on the above result which might be true, and which would be good to have, would be that the range of $Q^{(p)}$ can be pulled back to the quotient

$$\tilde{K}_*(X; \mathbf{Z}/p)/\{x^p : x \in \tilde{K}_*(X; \mathbf{Z}/p)\}.$$

Certainly the examples I have found do not seem to contradict this.

2. How to define $Q^{(p)}$

As with the case of homology, the best starting point is to consider a cyclic subgroup of order p, $\pi_p \subset \mathfrak{S}_p$, and look at the functor $K_*^{\pi_p}(X^p; \mathbf{Z}/p)$. First we need some notation. Let

$$\otimes^p(K_*(X; \mathbf{Z}/p))_{\pi_p} = H_0(\pi_p; \otimes^p(K_*(X; \mathbf{Z}/p)))$$

be the module of coinvariants with respect to the usual action of $\mathfrak{S}_p$, and so of π_p, on the p-fold tensor product. Let us also write J_{π_p} for $\tilde{K}_*(B\pi_p; \mathbf{Z}/p)$.

Theorem 3. <u>There is a natural short exact sequence</u> (on <u>Top</u>)

$$0 \to \otimes^p(K_*(X; \mathbf{Z}/p))_{\pi_p} \to K_*^{\pi_p}(X^p; \mathbf{Z}/p) \to J\pi_p \otimes K_*(X; \mathbf{Z}/p) \to 0.$$

<u>One consequence of Theorem 1, then, is that this sequence is not naturally split.</u>

The proof of Theorem 3 comes from the covering spectral sequence

$$H_*(\pi_p; K_*(X^p; \mathbf{Z}/p)) \Rightarrow K_*^{\pi_p}(X^p; \mathbf{Z}/p)$$

defined in the usual way. The E^2 term can be described in terms of generators as Dyer and Lashof did for homology; E^2_0 is the module of coinvariants (which maps into $K_*^{\pi_p}$, therefore, by the edge homomorphism) and E^2_q for $q > 0$ is all elements $e_q \otimes \overbrace{x \otimes \dots \otimes x}^{p} = e_q \otimes x^{(p)}$, where e_q is a q-cell in $B\pi_p$. The spectral sequence is a comodule over the Atiyah-Hirzebruch spectral sequence for $B\pi_p$. We find

$$d^{2p-1}(e_q \otimes x^{(p)}) = e_{q-2p+1} \otimes x^{(p)} \quad \text{for } q \text{ even}, \quad q \geq 2p$$

and E_q^{2p} vanishes except for $q = 0, 2, \ldots, 2p-2$. Hence $E^{2p} = E^\infty$. Now look at the secondary edge homomorphism. This maps

$$K_*(X; \mathbf{Z}/p) \to E_2^2 \subsetneq K_*^{\pi_p}(X^p; \mathbf{Z}/p)/E_0^\infty$$

by sending x to the class of $e_2 \otimes x^{(p)}$. By using the comodule structure over $K_*(B\pi_p; \mathbf{Z}/p)$ we can extend the above to the required isomorphism of $J\pi_p \otimes K_*(X; \mathbf{Z}/p)$ with $K_*^{\pi_p}(X^p; \mathbf{Z}/p)/E_0^\infty$ - and $E_0^\infty = E_0^2 = \otimes^p(K_*(X; \mathbf{Z}/p))_{\pi_p}$ is the module of coinvariants.

Next, we go over from π_p to $\mathfrak{S}_p$. There is one simple point to note here - that the normalizer $N(\pi_p)$ in $\mathfrak{S}_p$ permutes a set of generators of $J\pi_p$, so that our isomorphism above fits into a commutative diagram

$$\begin{array}{ccc} J\pi_p \otimes K_*(X; \mathbf{Z}/p) & \longrightarrow & K_*^{\pi_p}(X; \mathbf{Z}/p)/\otimes^p(K_*(X; \mathbf{Z}/p))_{\pi_p} \\ \downarrow & & \downarrow \\ K_*(X; \mathbf{Z}/p) & \xrightarrow{\phi_p} & K_*^{\mathfrak{S}_p}(X; \mathbf{Z}/p)(\otimes^p(K_*(X; \mathbf{Z}/p))_{\mathfrak{S}_p} \end{array}$$

And ϕ_p is the operation referred to in Theorem 2. To define $Q^{(p)}$, when X is an H^∞-space, first use a relative version of ϕ_p to map $\tilde{K}_*(X; \mathbf{Z}/p)$ into the direct summand

$$K_*^{\mathfrak{S}_p}((X, x_0)^p; \mathbf{Z}/p)/\otimes^p(\tilde{K}_*(X; \mathbf{Z}/p))_{\mathfrak{S}_p} \subset K_*^{\mathfrak{S}_p}(X^p; \mathbf{Z}/p)/\otimes^p(K_*(X; \mathbf{Z}/p))_{\mathfrak{S}_p}$$

Now apply μ_{p_*}. The composite

$$\otimes^p(K_*(X; \mathbf{Z}/p)) \to \otimes^p(K_*(X; \mathbf{Z}/p))_{\mathfrak{S}_p} \to K_*^{\mathfrak{S}_p}(X^p; \mathbf{Z}/p) \xrightarrow{\mu_{p_*}} K_*(X; \mathbf{Z}/p)$$

is just the p-fold Pontryagin product, hence the indeterminacy of $Q^{(p)} = \mu_{p_*} \circ \phi_p$ as defined above is contained in $(\tilde{K}_*(X; \mathbf{Z}/p))^p$.

I shall not prove the claim about the generators of $K_*(QS^0)$ here - it involves looking at the relation between the θ's as I defined them and the symmetric groups under the inclusion

$$B\mathfrak{S}_* = \amalg B\mathfrak{S}_n \to QS^0.$$

Essentially $K_*^{\mathfrak{S}_p}(B\mathfrak{S}_{p^r}; \mathbf{Z}/p)$ is the K-theory of $B(\mathfrak{S}_p \int \mathfrak{S}_{p^r})$ where $\mathfrak{S}_p \int \mathfrak{S}_{p^r}$ is the wreath product contained in $\mathfrak{S}_{p^{r+1}}$, and we have to identify $\theta_{p^{r+1}}$, inductively, as coming from this wreath product.

3. About Theorem 1

The counter-example X is the $(2p-1)$-skeleton of $(QS^0)_0$, which I shall call Y. We have

$$\tilde{H}_i(X; \mathbf{Z}/p) = \begin{cases} \mathbf{Z}/p & (i = 2p-3,\ 2p-2) \\ 0 & \text{otherwise} \end{cases}$$

and the Atiyah-Hirzebruch spectral sequence of $B\mathfrak{S}_p$ shows that we can choose a generator u of $\tilde{K}_0(X; \mathbf{Z}/p) = \mathbf{Z}/p$ such that $i_*(u) = \tilde{\theta}_p$. Then the Bockstein $\beta(u)$ generates $K_1(X; \mathbf{Z}/p)$.

Now let $\bar{f}: Y \to Y$ be the $(p+1)$st power map already referred to; then $\bar{f}$ restricts to $f: X \to X$, and f_* is the identity on $K_*(X; \mathbf{Z}/p)$.

Next, $\tilde{\theta}_{p^2}$ comes from $B\mathfrak{S}_p$, and an argument using wreath products (similar to the one I just referred to), plus use of filtrations in K_*, shows that there is a class, say v, in $K_*^{\mathfrak{S}_p}(X^p; \mathbf{Z}/p)$ which maps to $\tilde{\theta}_{p^2}$ under the composite

$$K_0^{\mathfrak{S}_p}(X^p; \mathbf{Z}/p) \xrightarrow{i_*} K_0^{\mathfrak{S}_p}(Y^p; \mathbf{Z}/p) \xrightarrow{\mu_{p*}} K_0(Y; \mathbf{Z}/p).$$

v will in fact be a representative of $\phi_p(u)$. I claim that $K_0^{\mathfrak{S}_p}(f^p; \mathbf{Z}/p)(v) \neq v$. The diagram

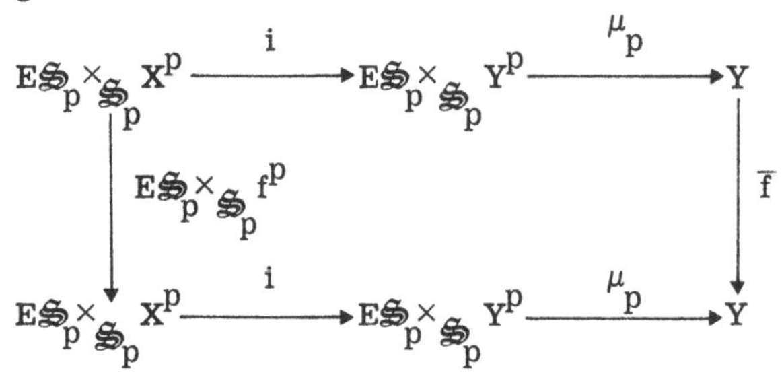

$$\begin{array}{ccccc} E\mathfrak{S}_p \times_{\mathfrak{S}_p} X^p & \xrightarrow{i} & E\mathfrak{S}_p \times_{\mathfrak{S}_p} Y^p & \xrightarrow{\mu_p} & Y \\ \downarrow{\scriptstyle E\mathfrak{S}_p \times_{\mathfrak{S}_p} f^p} & & & & \downarrow{\scriptstyle \bar{f}} \\ E\mathfrak{S}_p \times_{\mathfrak{S}_p} X^p & \xrightarrow{i} & E\mathfrak{S}_p \times_{\mathfrak{S}_p} Y^p & \xrightarrow{\mu_p} & Y \end{array}$$

is commutative, since $\bar{f}$ is an H^∞ map, and $\bar{f}_*(\tilde{\theta}_{p^2}) \neq \tilde{\theta}_{p^2}$ by Proposition 1. Hence

$$K_0(E\mathfrak{S}_p \times_{\mathfrak{S}_p} f^p)(v), \quad \text{or} \quad K_0^{\mathfrak{S}_p}(f^p; \mathbf{Z}/p)(v) \neq v$$

as I claimed.

4. Prospects

This work has made some progress, but we are still very far from knowing the Dyer-Lashof theory in K-theory as we do in homology - not to mention other theories. I have already suggested that it might be possible to improve the indeterminacy of $Q^{(p)}$. The example of QS^0 suggests also that we might be able to filter $K_*(X; \mathbf{Z}/p)$ and make $Q^{(p)}$ act in the associated graded ring. Among other questions I have not begun to look at seriously are

1. K-theory of QX - the method of Dyer and Lashof is no use here.
2. Cartan formula, Nishida relations and anything else nontrivial that might exist.

REFERENCES

1. E. Dyer and R. K. Lashof. Homology of iterated loop spaces. Amer. J. Math. 84 (1962), 35-88.
2. L. Hodgkin. The K-theory of QS^0. (To appear in Topology.)
3. J. P. May. Homology operations on infinite loop spaces (to appear).
4. G. Nishida. Cohomology operations in iterated loop spaces. Proc. Japan Acad. 44 (1968), 104-9.

King's College
London

HOMOTOPY HOMOMORPHISMS OF LIE GROUPS

J. R. HUBBUCK

Let $\mathcal{C}$ be the category whose objects are pairs $(\{X\}, BX)$ where $\{X\}$ is the homotopy class of a finite connected CW-complex X and BX is a complex with $\Omega BX \simeq X$. A morphism between objects BX and BY is a homotopy class of maps $f \in [BX, BY]$. This category is interesting because of the light it throws on the category of compact, connected Lie groups and homomorphisms. We recall a result of Milnor [10] which says that if Y is a countable, connected CW-complex, then ΩY has the homotopy type of a topological group, and it was once conjectured that if $(\{X\}, BX) \in \mathcal{C}$, then $X \simeq G$ a Lie group, but this latter is false [12]. However it appears that the only known objects of $\mathcal{C}$ occur when X has the genus of a Lie group [11], and no one has yet made explicit a multiplication which is not locally Lie, $(BX)_p \simeq (BG)_p$ for each p.

We investigate the connection between the morphisms in these two categories for objects common to both. The strongest conjecture that one might make is Problem 20 of [14], that the homotopy classes in Hom(G, H) correspond bijectively with [BG, BH], but Sullivan has shown that this is false. We consider the case when $G = H$ is a compact, simply connected, simple Lie group with its usual multiplication, so that Hom(G, G), apart from the trivial homomorphism, is Aut(G, G). The inner automorphisms are homotopic through inner automorphisms to the identity map, since G is connected. Outer automorphisms can be detected from symmetries in the Dynkin-Coxeter diagram of the associated Lie algebras and exist only for SU(n), $n > 2$; Spin(2k) and E_6. In these cases the quotient group of outer automorphisms is isomorphic to Z_2, except for Spin(8) when it is isomorphic to the symmetric group Σ_3, see section 33 of [6], and these are homotopically distinct.

Let $G = S^3$. The problem of determining $[BS^3, BS^3]$ has been considered by a number of people. First Berstein, and Cooke, Smith and

Stong in unpublished work in 1967 proved that the degree of any $f : BS^3 \to BS^3$ on the bottom cell S^4 was zero or the square of an integer, and results of Arkowitz and Curjel implied that if the degree was even, then it was divisible by 16. Later Adams† and Cooke showed that no even degrees were possible. D. Sullivan in Corollary 5.10 of [15] has shown that all remaining possibilities do occur.

More generally, if $i : G \to U$ is an inclusion of G in the infinite unitary group, I have heard a rumour that Sullivan has shown that when k is an integer prime to the order of the Weyl group of G, then there exists a map $f : BG \to BG$ such that the diagram

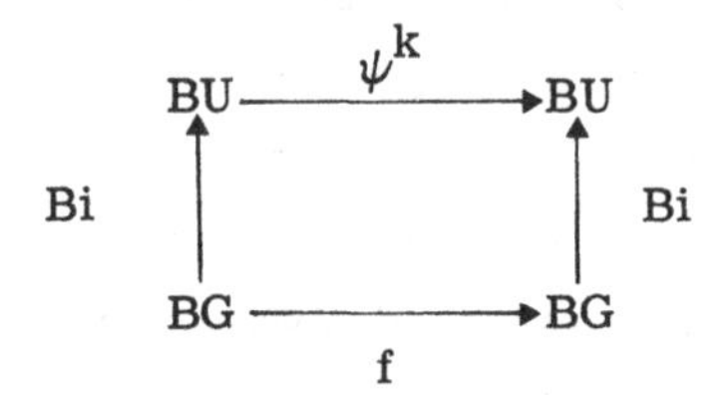

is homotopy commutative. A proof for $G = SU(n)$ with the standard inclusion is given as Corollary 5.11 of [15]. We seek a converse result. In the statements of the theorems which follow $f : BG \to BG$ is a given map. Assume that G is not the exceptional group G_2.

Theorem 1. There exists a non negative integer k and an outer automorphism $\tau : G \to G$ such that the following diagram is homotopy commutative,

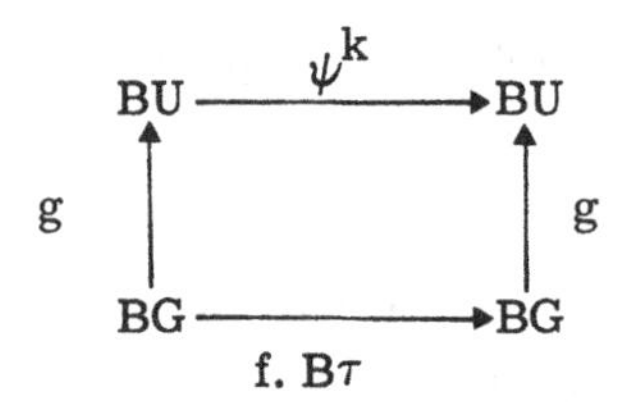

for all $g : BG \to BU$.

Remarks 1. The proof uses the known classification of the simple Lie groups. The proof for the exceptional groups is joint work with

---

† I am grateful to Professor Adams for showing his argument to me, and for several comments which he made on this and related papers.

Z. Mahmud.

2. BU may be replaced by BO.

3. If $G \neq Spin(4r)$ and we allow negative k, then the automorphism can be omitted.

4. The theorem holds equally for any $(\{X\}, BX)$ if $(BX)_p \simeq (BG)_p$ at each prime p, and $G \neq Spin(4r)$. A weaker result holds in the latter case.

There are two ways in which one might hope to strengthen this result.

1. Does the integer k determine the homotopy class of f?

2. Sullivan has conjectured for $SU(n)$ that ψ^k cannot be realized by $f : BG \to BG$ if k is not prime to the order of the Weyl group of G.

If this conjecture is correct for general G, then Theorem 1 holds for G_2.

We prove the theorem in an equivalent form. Recall that the rational cohomology ring $H^*(BG, Q)$ is a polynomial algebra on generators whose number equals the rank of G.

Theorem 2. *There exists a non negative integer* k *and an outer automorphism* $\tau : G \to G$ *such that if* u *is any class of* $H^*(BG, Q)$, *then* $(f.B\tau)^*u = k^q u$, *where* u *has degree* $2q$.

The Chern character $ch : K(BG) \to H^{even}(BG, Q)$ is monomorphic. Also if we use ch to identify $K(BG) \otimes Q$ with $H^{even}(BG, Q)$, then the action of ψ^k restricted to $H^{2q}(BG, Q)$ is multiplication by k^q. It follows easily that Theorems 1 and 2 are equivalent.

The homotopy groups $\pi_i(BG) = 0$ for $1 \leq i \leq 3$ and $\pi_4(BG) = Z$.

Corollary 3. $f_\# : \pi_4(BG) \to \pi_4(BG)$ *has degree* k^2 *for some integer* k.

Remarks on the proof of Theorem 2. We consider the restrictions placed on $f^* : H^*(BG, Q) \to H^*(BG, Q)$ implied by the requirement that the homomorphism $f^* : H^*(BG, Z_p) \to H^*(BG, Z_p)$ commutes with the action of the Steenrod algebra $\mathcal{A}(p)$, for each prime p. In principle the action of the Steenrod algebra can be calculated explicitly. In practice such an approach is too cumbersome and we prove a number of lemmas on homomorphisms of polynomial algebras over $\mathcal{A}(p)$. Details for most of the classical groups are in [7], and the rather different arguments needed for the exceptional Lie groups will appear in [8]. However these general arguments seem to require that the rank of the group is at least 5, and for groups of lower rank we resort to direct calculation. There are four groups of particular interest. These are Spin(8), because of the large group of outer automorphisms, F_4 which requires extensive calculations in the cohomology,† Sp(2) and G_2. Here we consider the simplest of these groups, Sp(2).

The ring $H^*(BSp(2), Z) \cong Z[x, y]$, a polynomial algebra on generators of dimensions 4 and 8 respectively. Let $f^*x = \alpha x$, $f^*y = \beta y + \gamma x^2$, where α, β and γ are integers. We consider the action of the cyclic reduced power P^1 on $H^*(BSp(2), Z_p)$ for odd primes p.

If $p = 2s - 1$,

$$P^1 x = a_0(p)x^s + a_1(p)x^{s-2}y + \dots$$

and if $p = 4t - 3$,

$$P^1 y = b_0(p)y^t + b_1(p)y^{t-1}x^2 + \dots$$

Of course x and y here stand for the reduction mod p of the corresponding integral classes. It is a classical result of number theory that there are an infinite number of primes of the form $4t - 3$ or $4u - 1$.

Lemma. (i) $a_0(p) = 2 \ (p)$, <u>and if</u> s <u>is even</u>, $a_{s/2}(p) = (-1)^{s/2}4 \ (p)$.
(ii) $b_0(p) = (-1)^{t-1}4 \ (p)$.

Proof. Let T be the standard maximal torus in Sp(2). Then $H^*(BSp(2), Z)$ can be identified with the polynomial subalgebra of

† This result is due to Z. Mahmud.

$H^*(BT, Z) \cong Z[t_1, t_2]$ generated by $x = t_1^2 + t_2^2$ and $y = t_1^2 t_2^2$. Thus we perform the calculations in $H^*(BT, Z_p)$. We give the details for $a_0(p)$. Let $\phi^q = t_1^{2q} + t_2^{2q}$, and recall that for $q > 2$, $\phi^q = \phi^{q-1}.x - \phi^{q-2}.y$ and $\phi^2 = x^2 - 2y$. Now $P^1(\phi^1) = 2\phi^s$. The recurrence relation implies that when ϕ^s is expressed as a polynomial in x and y, then the coefficient of x^s is 1. Thus $a_0(p) = 2$. The calculations for $a_{s/2}(p)$ and $b_0(p)$ are similar.

We consider the equality

$$(*) \qquad P^1 f^* x = f^* P^1 x.$$

Assume that s is even. Equating coefficients of $y^{s/2}$ on either side of $(*)$ implies that

$$(1) \qquad a_{s/2}.\alpha = a_{s/2}.\beta^{s/2} \quad (p).$$

As $a_{s/2} \neq 0$ (p), we have $\alpha = \beta^{s/2}$ (p) and so $\alpha^4 = \beta^{p+1}$ (p). If p is not a divisor of β, $\beta^{p-1} = 1$ (p) and so $\alpha^4 = \beta^2$ (p). As p may be chosen to be arbitrarily large, it follows that $\alpha^4 = \beta^2$. There are now two cases to consider separately; either $\beta = \alpha^2$ or $\beta = -\alpha^2$.

Case 1. Assume that $\beta = \alpha^2$. Again assuming that s is even, we equate the coefficients of $y^{(s-2)/2}.x^2$ on either side of $(*)$.

$$\alpha.a_{(s-2)/2} = \alpha^s.a_{(s-2)/2} + a_{s/2}.\alpha^{2s-2}.s/2.\gamma \quad (p),$$

which when combined with (1) gives

$$a_{s/2}.\alpha^{(p-1)/2}.-2^2.\gamma = 0 \quad (p).$$

Therefore if $\alpha \neq 0$ and p is chosen large, since $a_{s/2} \neq 0$ (p), we deduce that $\gamma = 0$.

If $\alpha = 0$, we consider the coefficients of x^s in $(*)$, and deduce that $0 = a_{s/2}.\gamma^{s/2}$ (p), which again leads to the conclusion that $\gamma = 0$.

Next we quote a result from prime number theory which is a consequence of a famous theorem of Dirichlet.

Lemma. *If* n *is an integer and* $n^{(p+1)/2} = n \quad (p)$ *for all primes* p, *then* $n = m^2$ *for some integer* m.

We use (*) again, this time for each odd prime, and consider coefficients of x^s. Since $a_0 \neq 0 \quad (p)$ by the lemma, we have $\alpha = \alpha^{(p+1)/2} \quad (p)$ for all odd primes and therefore for all primes and therefore $\alpha = k^2$. Recapitulating, we have established that $\alpha = k^2$, $\beta = k^4$ and $\gamma = 0$.

Case 2. This occurs when $\beta = -\alpha^2$. Again equating coefficients of $y^{s/2}$ on either side of (*) leads to the conclusion that $\alpha = (-1)^{s/2}\alpha^{(p+1)/2} \quad (p)$. Thus

$$\alpha^{(p+1)/2} = \alpha \quad (p) \text{ if } p = -1 \quad (8); \quad \alpha^{(p+1)/2} = -\alpha \quad (p) \text{ if } p = 3 \quad (8).$$

Now we apply a similar argument to the coefficients of y^t on either side of $P^1 f^* y = f^* P^1 y$ when $p = 4t - 3$. This leads to $-b_0 . \alpha^2 = (-1)^t b_0 . \alpha^{2t} \quad (p)$, and as $b_0 \neq 0 \quad (p)$,

$$\alpha^{(p+1)/2} = \alpha \quad (p) \text{ if } p = 1 \quad (8); \quad \alpha^{(p+1)/2} = -\alpha \quad (p) \text{ if } p = -3 \quad (8).$$

One solution for α of all four congruence relations is 2, and so the general solution is $\alpha = 2k^2$.

We could proceed further and show that $\alpha = 2k^2$, $\beta = -4k^4$ and $\gamma = k^4$ (and nothing more using primary operations), but all we require now is the observation that if f comes under case 2, then the composition of f with itself comes under case 1. Thus we have established,

Proposition 4. *If there does not exist an integer* k *such that* $f^* u = k^q u$ *for each (homogeneous) class* u *in* $H^*(BSp(2), Z)$ *of degree* $2q$, *then there is an integer* k *such that* $(f.f)^* u = 2^q k^{2q} u$.

The proof of Theorem 2 for $Sp(2)$ is now completed by the next proposition.

Proposition 5. *Suppose that* G *has a complex irreducible representation which is also quaternionic. If* $f : BG \to BG$ *realizes* ψ^k *in the sense of Theorem 1, where* k *is non zero, then* k *is an odd*

integer.

The groups which satisfy the hypotheses of Proposition 5 are SU(4n+2), Sp(n), Spin(8n+4) and Spin(8n±3), see Chapter 7 of [1] and Chapter 13 of [9].

Proof of Proposition 5. We consider the commutative diagram

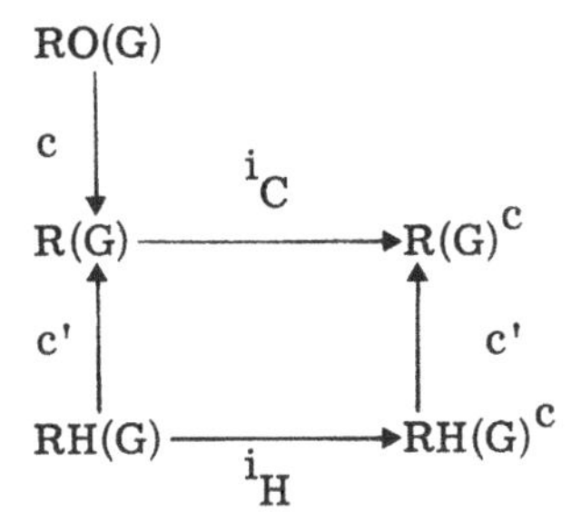

R(G) is the complex representation ring of G and $R(G)^c$ is its completion with respect to the augmentation topology of R(G). A theorem of Atiyah and Hirzebruch (section 4. 8 of [4]) implies that $R(G)^c$ can be identified with the complex K-group K(BG). The analogous result for quaternionic K-theory is due to Anderson [2]. RH(G) is the quaternionic representation group of G and $RH(G)^c$, its completion with respect to a suitable topology, can be identified with KH(BG). Anderson defined this topology by using the CW-topology on KH(BG), but equivalently we may use the IO(G)-adic topology, where IO(G) is the augmentation ideal of the real representation ring RO(G) [5]. All homomorphisms in the diagram, the inclusions i_C and i_H and the complexifications c and c', are monic, and so we regard all groups as embedded in $R(G)^c \cong K(BG)$.

Let $x \in RH(G)$ be a quaternionic, complex irreducible representation. Since $x \in KH(BG)$, $f^! x \in KH(BG)$ by the naturality of c'. Since $x \in R(G)$, $\psi^k x \in R(G)$. But $R(G) \cap KH(BG) = RH(G)$ and so $f^! x = \psi^k x \in RH(G)$. Now assume that k is a (non zero) even integer, so that $\psi^k x \in RO(G)$, see remark 3. 64 of [1]. We shall obtain a contradiction. Now $RO(G) \cap RH(G) = (1 + t)R(G)$ where t is conjugation and so $\psi^k x = (1 + t)y$ for some $y \in R(G)$. Recall that $R(G) = R(T)^W$, the ring of invariants under the action of the Weyl group of a maximal torus of G, and $R(T) = Z[x_i, x_i^{-1}]$, $1 \le i \le \text{rank } G$, where the x_i are one dimensional representations. Suppose first that t is the identity. This will be true unless $G = SU(4n+2)$. Thus $\psi^k x = 2y$. But $\psi^k(p(x_i, x_i^{-1})) = p(x_i^k, x_i^{-k})$ for any polynomial in $R(T)$ and $p(x_i^k, x_i^{-k})$ is divisible by 2 only if $p(x_i, x_i^{-1})$ is divisible by 2. So in $R(T)$, x is divisible by 2. However if $2u \in R(T)$ is invariant under the action of the Weyl group, so is u, and so x is divisible by 2 in $R(G)$, which is false since x is irreducible.

It remains to prove the proposition for $SU(4n+2)$. In standard notation the unique quaternionic irreducible representation of $SU(4n+2)$ is λ_{2n+1} which restricts to $\sigma_{2n+1}(x_1, \ldots, x_{4n+2})$ on T, where σ_i is the elementary symmetric sum of the x_i. Thus in $R(T)$, $\psi^k(\sigma_{2n+1}) = \sigma_{2n+1}(x_1^k, \ldots, x_{2n+2}^k) = (1+t)y$. As $(1+t)y = (1+t)ty$, we may assume without loss of generality that the coefficient of $x_1^k \ldots x_{2n+1}^k$ in y is non zero mod 2. But the Weyl group is the symmetric group and acts by permuting the x_i. Thus $y = \sigma_{2n+1}(x_1^k, \ldots, x_{4n+2}^k) + W \bmod 2$, where W is a symmetric expression in the x_i not containing any terms of the form $x_{i_1}^k \ldots x_{i_{2n+1}}^k$, where the subscripts i_j are all distinct. But clearly $(1+t)y$ does not equal $\sigma_{2n+1}(x_1^k, \ldots, x_{4n+2}^k)$, and we have obtained a contradiction.

REFERENCES

1. J. F. Adams. Lectures on Lie groups (Benjamin, 1969).
2. D. W. Anderson. The real K-theory of classifying spaces. Proc. Nat. Acad. Sci. U.S.A. 51 (1964), 634-6.
3. M. F. Atiyah. K-theory (Benjamin, 1967).
4. M. F. Atiyah and F. Hirzebruch. Vector bundles and homogeneous spaces. Proc. of Symposia in Pure Mathematics 3, Differential Geometry, Amer. Math. Soc. (1961), 7-38.
5. M. F. Atiyah and G. B. Segal. Equivariant K-theory and completion. J. of Differential Geometry 3 (1969), 1-18.
6. Freudenthal and de Vries. Linear Lie groups (Academic Press, 1969).
7. J. R. Hubbuck. Mapping degrees for classifying spaces, 1. Preprint.

8. J. R. Hubbuck and Z. Mahmud.

9. D. Husemoller, Fibre Bundles (McGraw-Hill, 1966).

10. J. Milnor. Construction of Universal bundles, 1 . Ann. of Math. (2) 63 (1956), 272-84.

11. G. Mislin. The Genus of an H-space , Symposium on Algebraic Topology. Lecture Notes in Mathematics, 249 (Springer-Verlag, 1971).

12. J. D. S. Stasheff. Manifolds of the homotopy type of (non-Lie) groups . Bull. Amer. Math. Soc. 75 (1969), 998-1000.

13. J. D. S. Stasheff (ed.). Problems proposed at the conference. Conference on Algebraic Topology, University of Illinois at Chicago Circle, 1968, 288-93.

14. J. D. S. Stasheff (ed.). H-space problems, H-spaces Neuchatel (Suisse) Aout 1970. Lecture Notes in Mathematics (Springer-Verlag, 196).

15. D. Sullivan. Geometric Topology, 1 . Notes M. I. T. (1970).

Gonville and Caius College, Cambridge
Magdalen College, Oxford

ON SPHERICAL FIBER BUNDLES AND THEIR PL REDUCTIONS

IB MADSEN and R. JAMES MILGRAM†

Twenty years ago, Borel, Bott, Hirzebruch, Milnor, and Thom, among others, studied the structure of the classifying spaces for the orthogonal and unitary groups. From their work, it became clear that the classifying spaces B_{PL}, B_{TOP}, and B_G (B_G is the classifying space for fiber homotopy sphere bundles [14], [20]) contained the answers to many of the problems they raised.

The last ten years have seen a concerted effort to understand these spaces, and the path has been highlighted by several beautiful results: Sullivan's work on G/PL and related spaces leading to the Hauptvermutung for 4-connected manifolds ([18], [21]), Novikov's work on the invariance of the rational Pontrjagin classes ([15]), the work of Kirby-Siebenmann and Lashof-Rothenberg on G/TOP and the triangulation theorem ([7], [9]), and the work of Quillen-Sullivan on the Adams conjecture ([16], [22]).

Recently, in joint work with Brumfiel, we have determined the mod 2 cohomology of B_{PL} and B_{TOP} ([3]). This of course gave the algebraic determination of the unoriented PL-bordism ring and, except in dimension 4, the topological bordism ring.

Here we almost complete the analysis of the structure of $H^*(B_{PL})$ at the prime 2. In particular, at the prime 2, we determine the obstructions to reducing the structure 'group' of a fiber homotopy sphere bundle to TOP or PL. As an application, using the Browder-Novikov theorem, these obstructions determine explicit conditions on a simple-connected Poincare-duality space, which imply that it has the (2-local) homotopy type of a topological or PL-manifold. Also, the result gives the (2-local) structure of the oriented PL-bordism ring $\Omega^{PL}{}_*(\ , Z_{(2)})$ and, except

---

† This work partially supported by NSF Grant GP 29696 A1.

in dimension 4, $\Omega^{TOP}{}_*(\ ,Z_{(2)})$, where $Z_{(2)}$ denotes the integers localized at the prime 2 [24].

The method is to look at the fibrations

$$G/PL \to B_{PL} \to B_G \xrightarrow{\pi} B_{(G/PL)}.$$

The (2-local) structure of B_G is known from [14] and [10]. We next prove

Theorem A. <u>At the prime</u> 2, $B_{(G/PL)}$ <u>is a product</u>

$$E_3 \times \prod_{i=2}^{\infty} K(Z_{(2)}, 4i+1) \times K(Z_2, 4i-1),$$

<u>while</u> $B_{G/TOP}$ <u>is simply a product of Eilenberg-MacLane spaces.</u>

Theorem A has two immediate corollaries.

Corollary B. <u>At the prime</u> 2, <u>the obstructions to the existence of a PL-bundle structure on a fiber homotopy sphere bundle are:</u> (1) <u>a secondary characteristic class in dimension</u> 5, <u>and</u> (2) <u>in odd dimensions,</u> $n \geq 5$, <u>ordinary</u> $Z_{(2)}$- <u>or</u> Z_2-<u>characteristic classes.</u>

Corollary C. $B_{(B_{G/PL})} = E_{3,1} \times \prod_{i=2}^{\infty} K(Z_{(2)}, 4i+2) \times K(Z_2, 4i).$

Corollary C implies that the fundamental classes in $H^*(B_{G/PL})$ may be taken as primitives.

The next step is to determine the map π^* (actually, this is also the last step, since A implies that standard spectral sequence techniques can now be used to obtain $H^*(B_{PL}, Z_{(2)})$). π factors through the composite

$$B_G \xrightarrow{\rho} B_{G/O} \xrightarrow{\tau} B_{G/PL}.$$

Using the ideas of [14], $H^*(B_{G/O})$ is easily computed and the map ρ^* is unambiguously defined. Thus the evaluation of π^* reduces to the evaluation of τ^*, but, on examining the suspension diagram

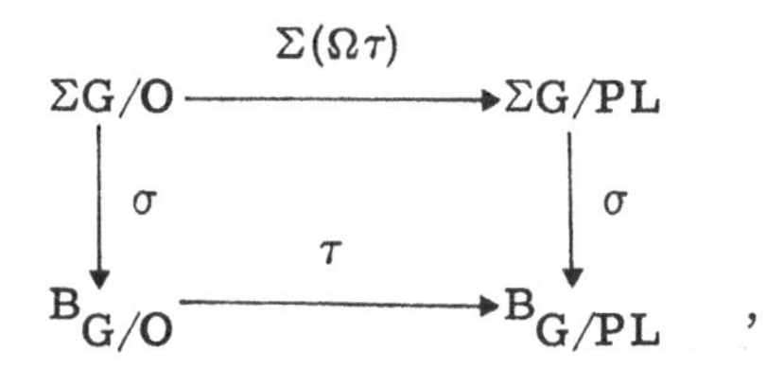

we obtain that τ^* is determined by $(\Omega\tau)^*$. At this stage, we use the proof of the Adams conjecture to obtain a map

$$B_{SO} \xrightarrow{\mathcal{L}} G/O \overset{\lambda}{\dashrightarrow} B_O ,$$

so $\lambda \circ \mathcal{L} = \psi^3 - 1$ at the prime 2. Moreover, $\mathcal{L}^*$ and $(\Omega\tau \cdot \mathcal{L})^*$ can be completely determined. Putting this result together with a slight extension of the results of [3], we determine π^*.

To describe the main results, consider the Hopf algebra

$$\mathcal{H} = P(p_1, \ldots, p_n \ldots) \otimes E(b_1, \ldots, b_n \ldots) \quad (\text{over } Z_{(2)})$$

with $\psi(p_i) = \Sigma^i_{r=0} p_r \otimes p_{i-r}$, $\psi(b_i) = \Sigma^i_{r=1}(b_r \otimes p_{i-r} + p_{i-r} \otimes b_r)$, dimension $p_i = 4i$, dimension $b_i = 4i + 1$. Introduce a derivation δ by setting $\delta p_i = 8b_i$. In each dimension $4i + 1$, there is a primitive in $\mathcal{H}$ of the form b_i + (decomposables). The first few are b_1, $b_2 - b_1p_1$, $b_3 - p_1b_2 - (p_2 - p_1^2)b_1 = b_3 - p_1q_2 - p_2q_1$. Indeed, we have inductively

$$\text{(D)} \qquad q_i = b_i - \Sigma q_r p_{i-r}.$$

Let s_i be the primitive in $\mathcal{H}$ of dimension $4i$. s_i is given by the Newton formula

$$s_i = p_1 s_{i-1} - p_2 s_{i-2} + \ldots + (-1)^i i(p_i)$$

and $\delta(s_i) = 8iq_i$. Let $\nu(i)$ be the largest power of 2 dividing i. Then we have

Lemma E. $H_*(\mathcal{H}, \delta)$ <u>has non-zero primitive classes</u> $\{q_i\}$ <u>in dimension</u> $4i + 1$, <u>and the</u> 2-<u>order of</u> $\{q_i\}$ <u>is</u> $8 \cdot \nu(i)$.

The next result, together with universal facts about modules over the Dyer-Lashof algebra ([10]), completely determines the $Z_{(2)}$-cohomology structure of $H^*(B_G)$.

Lemma F. There is a (2-local) injection $j : H_*(\mathcal{K}, \delta) \to H^*(B_G, Z_{(2)})$ so that, on taking Z_8-coefficients, the composite $H_*(\mathcal{K}, Z_8) \to H^*(B_G) \to H^*(B_O, Z_8)$ takes $\{p_i\}$ to the mod 8 reduction of the ith Pontrjagin class. Moreover, there is a Hopf algebra $\mathcal{K}$ with derivation δ', so $H^*(B_G, Z_{(2)}) \cong H^*(\mathcal{K} \otimes \mathcal{K}, \delta \otimes 1 + \varepsilon \otimes \delta')$, and j is the natural inclusion.

In Milgram ([23]), a universal surgery class $K_{4*} \in H^{4*}(G/TOP, Z_{(2)})$ was constructed. The class K_{4*} is not primitive,

$$K_{4n} \to 1 \otimes K_{4n} + 8\Sigma K_{4i} \otimes K_{4j} + K_{4n} \otimes 1.$$

It is easy, however, to construct a primitive class

$$k_{4n} = K_{4n} + 4\cdot \text{(decomposable elements)},$$

where the decomposable elements are a polynomial in K_{4j}, $j < n$. Here is our main result ($\pi : B_G \xrightarrow{\rho} B_{G/O} \xrightarrow{\tau} B_{G/TOP}$).

Theorem G. There is a primitive graded class $k_{4*+1} \in H^{4*+1}(B_{G/TOP}; Z_{(2)})$ with the following properties:

(i) $(\Omega\tau)^*(\sigma k_{4*+1} - k_{4*}) = 0$ in $H^{4*}(G/O; Z_{(2)})$,

(ii) $\pi^*(k_{4i+1}) = 2^{\alpha(i)-1} j(q_i)$, $i \geq 1$.

(Here $\alpha(i)$ is the number of ones in the dyadic expansion of i.)

The following corollary shows that the surgery formula for tangential normal maps reduces considerably.

Corollary H. $(\Omega\pi)^*(K_{4n}) = 2^{\alpha(n)-1}\cdot Y_{4n}$, where $Y_{4n} \in H^{4n}(G; Z_{(2)})$ is a class of order 4.

Corollary J. There are primitive classes $\theta_{i,i} \in H^{2^{i+1}-1}(B_G; Z_2)$ so that a homotopy sphere bundle ξ on a finite complex X admits a *reduction to a topological sphere bundle* (2-*locally*) *if and only if*

(1) $\theta_{i,i}(\xi)$ *are all* 0, *and*

(2) $2^{\alpha(i)-1} j(q_i)(\xi) = 0.$

Remark K. Recent work of D. Ravenel constructs explicit characteristic classes λ_i which differ from the $\theta_{i,i}$ above only by decomposables. Moreover, his λ_i would be exactly our $\theta_{i,i}$ if they were known to be primitive! Mahowald has shown that, if X is a Poincare-duality space with vanishing Stiefel-Whitney classes, and if ξ is the Spivak normal spherical fibration, then $\theta_{i,i}(\xi)$ is the secondary Wu class associated to the Adams operation $\Phi_{i,i} : H^*(X; Z_2) \to H^*(X; Z_2)$. Here $\theta_{i,i}$ is the secondary operation based on the relation

$$\sum_{0 \leq j \leq i} \chi(Sq^{2^{i+1}-2^j})\chi(Sq^{2^j}) = 0.$$

It is interesting to compare these results with the 'transversality' results of Levitt-Quinn and Brumfiel-Morgan ([17], [4]). Recall that they have used and developed Levitt's ideas on measuring the obstructions to transversality for maps of Poincare-duality spaces in order to construct a fibering

$$B_{STOP} \to B_{SG} \xrightarrow{\mathcal{A}} \mathcal{K},$$

where $\mathcal{K} \simeq B_{(G/TOP)}$. At the prime 2, Brumfiel and Morgan calculate $\mathcal{A}^*$ with Z_2- and Z_4-coefficients, while showing that $8\mathcal{A}^*(K_{4i+1}) \equiv 0$. It was natural to conjecture that $\mathcal{A}$ and j are in some sense the same map, but from G we find

Corollary L. *There is no* (2-*local*) *homotopy equivalence*

$$\mu : \mathcal{K} \to B_{G/TOP}$$

for which $\mu \circ \mathcal{A}$ j.

Corollary L follows because there is no homotopy equivalence $k : B_{G/TOP} \to B_{G/TOP}$, so $j^*k^*(K_{4i+1})$ has 2-order dividing 8 for all i. However, in a certain sense, the (mod 8) reductions of the obstructions are all that matter, as we see from

Corollary M. Suppose $f : X \to B_{SG}$ lifts to B_{STOP} on the 4n-skeleton of X. Then f lifts to B_{STOP} (2-locally) on X_{4n+2} if and only if $f^*(L_{4n+1}) = 0$ where $L = 2^{\alpha(n)-1}y$, with y of order 8.

Finally, it is routine but very messy to calculate $H^*(B_{STOP}, Z_{(2)})$ and $H^*(B_{SPL}, Z_{(2)})$, using for example Serre spectral sequence techniques and the complete knowledge of the universal Serre sequences (on the chain level) for the loop path fibering

$$\Omega K(\pi, n) \to E \to K(\pi, n).$$

The proofs of most of the results above follow from [3] after we have proved A. We do this in the next two sections. The third section calculates the composite map $(\Omega\tau \cdot \mathcal{L})^* : H^*(G/PL) \otimes Q \to H^*(BS_O) \otimes Q$, and justifies the coefficient $2^{\alpha(i)-1}$ in Theorem G. Finally, in Section 4, we complete the determination of $(\Omega\tau)^*$. To this end, one splits the homology of G/O in two parts, $H_*(G/O) \approx H_*(BSO) \otimes Y$. The subalgebra Y of $H_*(G/O)$ is abstractly isomorphic to $H_*(\text{Cok } J)$. The primitive class $k_{4i} \in H^{4i}(G/PL)$ evaluates zero on Y, and this, together with the results of Section 3, gives Theorem G.

Warning. The classes $\tau^*(k_{4i}) \in H^*(G/O; Z_{(2)})$ do not vanish in $H^*(\text{Cok } J; Z_{(2)})$; e.g. $\tau^*(k_{12})$ restricts to a class of order 2 in $H^{12}(\text{Cok } J; Z_{(2)})$.

We would like to thank our collaborator Gregory Brumfiel for several illuminating conversations, comments, and examples, which unerringly pointed the way when the work seemed mired in incredibly messy case-by-case calculations.

§1. The Mahowald orientation

Consider the space $\Omega^n S^{n+1}$. Its cohomology structure has been completely determined in [1], [12]. In particular, we have

Lemma 1.1. (a) $H^*(\Omega^n S^{n+1}, Z_2) = E(\lambda_1, \ldots, \lambda_I \ldots)$, _an exterior algebra on stated generators._

(b) $H_*(\Omega^2 S^3, Z_2) = P(e_1, Q_1 e_1, Q_1 Q_1(e_1), \ldots)$, _one generator in each dimension of the form_ $2^i - 1$.

Consider the non-trivial map

$$S^1 \xrightarrow{\eta} B_0 .$$

Since B_O is an infinite loop space, corresponding to η are maps $\sigma^i(\eta) : S^{i+1} \to B^{(i)}_O$ (where $\Omega^i B^{(i)}_O = B_O$). In particular, corresponding to $\sigma^2(\eta) : S^3 \to B^{(2)}_O$, on looping twice we obtain the diagram

(1.2) 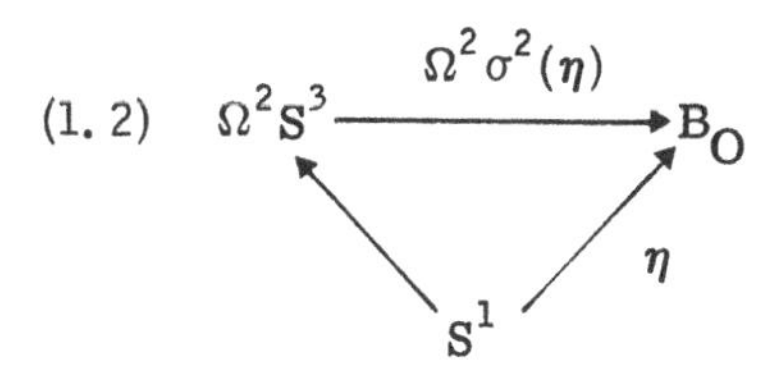

and we have

Theorem 1.3 (M. Mahowald). _Let_ γ _be the universal bundle over_ B_O, _and_ $\gamma = \Omega^2\sigma^2(\eta)^!(\gamma)$. _Then at the prime_ 2, $M(\bar{\gamma}) = K(Z_2, 0)$, _the Eilenberg-MacLane spectrum._

Proof. Let $H_*(B_O, Z_2)$ be written as $P(\bar{e}_1, \bar{e}_2, \ldots, \bar{e}_n \ldots)$, where $\bar{e}_n$ is dual to ω_1^n. Let e be the generator in $H_1(S^1, Z_2)$. Then $\eta_*(e) = \bar{e}_1$. Hence from (1.2), $\Omega^2\sigma^2(\eta)_*(e_1) = \bar{e}_1$. Now by Kochman's result ([8], Theorem 41), $Q_1(\bar{e}_1) = (\bar{e}_3)$, $Q_1(\bar{e}_3) = \bar{e}_7 \ldots \underbrace{Q_1 \ldots Q_1}_{i}(\bar{e}_1) = \bar{e}_{2^{i+1}-1}$.

Thus, since Ω^2 of any map commutes with the action of Q_1, the homology map is determined.

We now pass to Thom spaces. Recall that $\mathcal{A}(2)^* = P(\xi_1, \xi_3, \xi_7, \ldots, \xi_{2^i-1} \ldots)$, and let q_i be the primitive in $\mathcal{A}(2)$ dual to ξ_{2^i-1}. Then $Sq^1 = q_0$, $[Sq^2, Sq^1] = q_1, \ldots, q_i = [Sq^{2^i}, q_{i-1}]$.

Lemma 1.4. $\langle q_i U, \bar{e}^*_{2^{i+1}-1} U_* \rangle = 1$.

Proof. By induction: notice that $q_i U = q_{i-1}(\omega_1^{2^i} U) + Sq^2(\omega_1^{2^i - 1} U + TU)$, where $T \subset I(\omega_2, \omega_3, \ldots, \omega_n)$. But $\mathcal{A}(2) I(\omega_2, \ldots, \omega_n \ldots) \subset I(\omega_2, \ldots, \omega_n)$. Hence, mod I, $q_i U = Sq^{2^i} \omega_1^{2^i - 1} U = \omega_1^{2^{i+1}-1} U$, and 1.4 follows.

Next, notice that, if we define a map $\phi : \mathcal{A}(2) \to H^*(M(\gamma))$ by $\alpha \to \alpha U$, then using the multiplication in B_0 to give $M(\gamma) \wedge M(\gamma) \to M(\gamma)$, the diagram

$$
(1.5) \qquad \begin{array}{ccc}
\mathcal{A}(2) & \xrightarrow{\Delta} & \mathcal{A}(2) \otimes \mathcal{A}(2) \\
\downarrow \phi & & \downarrow \phi \otimes \phi \\
H^*(M(\gamma)) & \longrightarrow & H^*(M(\gamma)) \wedge M(\gamma) \\
\downarrow & & \downarrow \\
H^*(M(\bar{\gamma})) & \longrightarrow & H^*(M(\bar{\gamma}) \wedge M(\bar{\gamma}))
\end{array}
$$

commutes. This implies that the dual map

$$\phi_* : H_*(M(\bar{\gamma})) \to \mathcal{A}(2)^*$$

is a map of algebras. Lemma 1.4 now shows that $\phi_*(\underbrace{Q_1 \ldots Q_1}_{i}(e_1)U_*) = \xi_{2^{i+1}-1} + d$, where $d \in I(\xi_1 \ldots \xi_{2^i - 1})$. From this, it follows that ϕ_* and hence ϕ are isomorphisms, and this gives 1.3.

Generalizing (1.2) slightly, we have the diagram

$$(1.6) \qquad S^1 \to \Omega^2 S^3 \to \Omega^3 S^4 \to \Omega^4 S^5 \to \ldots \to \Omega^n S^{n-1} \to B_0.$$

Let $\bar{\gamma}_r = \Omega^r \Sigma^r(\eta)^!(\gamma)$. Then we have the Thom spaces $M(\bar{\gamma}_r)$ and the maps

$$(1.7) \qquad M(\bar{\gamma}) \to M(\bar{\gamma}_3) \to M(\bar{\gamma}_4) \to \ldots,$$

and, by [2], each $M(\bar{\gamma}_i)$ is a wedge of Eilenberg-MacLane spaces. Moreover, the map

$$M(\bar{\gamma}_i) \wedge M(\bar{\gamma}_i) \to M(\bar{\gamma}_i)$$

induced by Whitney bundle sum is the result of the H-map loop sum: $\Omega^i S^{i+1} \times \Omega^i S^{i+1} \to \Omega^i S^{i+1}$, which has more and more structure as $i \to \infty$.

Definition 1.8. A differentiable manifold M^n has a Mahowald orientation if the classifying map for its normal bundle

$$\nu : M^n \to B_0$$

factors through $\Omega^r \Sigma^r(\eta)$ for some r. We say it has a primitive Mahowald orientation if it factors through $\Omega^2 \Sigma^2(\eta)$.

Theorem 1.9. Let M^n have a Mahowald orientation. Then if ω is any positive-dimensional characteristic class of $\nu(M^n)$, we have $\omega^2 = 0$. In particular, $V^2(M^n) = 1$, where V is the total Wu class of M. Moreover, M^n is a Z_2-manifold. (Since $x^2 = 0$ for all $x \in H^*(\Omega^r S^{r+1}, Z_2)$, and e_1^* is an integral class, so $V_1(M)$ is always the restriction of an integral class.)

Theorem 1.10. Let $x \in H_n(X; Z_2)$. Then x admits a realization by a manifold and map (M^n, f) with a primitive Mahowald orientation. Moreover, up to Mahowald-oriented bordism, (M^n, f) is unique.

Proof. Let $\mathfrak{M}_*(X)$ be the bordism of X with respect to primitively Mahowald-oriented manifolds. Then $\mathfrak{M}_*(X) \cong \pi_*(X \wedge M(\bar{\gamma})) \cong H_*(X; Z_2)$ ([5]), and the result follows.

Remark 1.11. A good way of regarding the Mahowald orientation is as an explicit inverse of the Thom class map $M(\bar{\gamma}) \xrightarrow{U} K(Z_2, 0)$. But the miracle is that it actually is a Thom space, and hence satisfies transversality.

§2. The proof of Theorem A

The strategy is to calculate in the Eilenberg-Moore spectral sequence passing from $\mathrm{Tor}_{H_*(G/PL,\, Z_2)}(Z_2, Z_2)$ to $H_*(B_{G/PL}, Z_2)$. Of course, our attack must be made using manifolds. This suggests the use of the Eilenberg-Moore spectral sequence passing from

(2.1) $\mathrm{Tor}_{\Omega_*(G/PL,\ Z_2)}(\Omega_*(pt,\ Z_2),\ \Omega_*(pt,\ Z_2))$

to $\Omega_*(B_{G/PL};\ Z_2)$. There is a Hurewicz map of spectral sequences, which at the E^2-level is the change of rings map

(2.2) $\tilde{h} : \mathrm{Tor}_{\Omega_*(G/PL,\ Z_2)}(\Omega_*(pt,\ Z_2),\ \Omega_*(pt,\ Z_2)) \to \mathrm{Tor}_{H_*(G/PL,\ Z_2)}(Z_2,\ Z_2)$

induced by the Hurewicz map $h : \Omega_*(G/PL,\ Z_2) \to H_*(G/PL,\ Z_2)$ and the augmentation $h : \Omega_*(pt) \to H_*(pt)$. On the other hand, the Mahowald orientation induces a ring map $H_*(G/PL,\ Z_2) \to \Omega_*(G/PL,\ Z_2)$, which in turn induces a map of Eilenberg-Moore spectral sequences, which at E^2 is

$\tau : \mathrm{Tor}_{H_*(G/PL,\ Z_2)}(Z_2; Z_2) \to \mathrm{Tor}_{\Omega_*(G/PL,\ Z_2)}(\Omega_*(pt,\ Z_2),\ \Omega_*(pt,\ Z_2))$,

and $h \circ \tau = 1$. Moreover, this is true at all levels, and, in fact, the differentials in the bordism spectral sequence are completely determined by the differentials in the homology sequence from τ.

Generally speaking, any differential going to filtration 1 in these Eilenberg-Moore spectral sequences is determined by a matric Massey product being defined and non-zero in $H_*(G/PL,\ Z_2)$ or $\Omega_*(G/PL,\ Z_2)$. Indeed, if we have $x \in \langle A_1, \ldots, A_n \rangle$ and x not contained in any smaller product, then $d_{n-1}(|A_1| \ldots |A_n|) = \{x\}$ represents a non-zero differential.

Next, we observe that the Eilenberg-Moore spectral sequences are sequences of differential Hopf algebras, by results of A. Clark, and it is easy to see that, if any differentials are non-zero, the first such is d_{2^i-1} for some i, and there must be an element $y = |x_1| \ldots |x_{2^i}|$ with $d_{2^i-1}(y) \neq 0$. Moreover, except in certain cases coming from the peculiarities of the space $E_2 \subset G/PL$, these x_i are all equal to a single element $x \in H_*(G/PL,\ Z_2)$ (the remaining cases give terms $|x|y|x|y| \ldots |x|y|$).

Lemma 2.3. Suppose $x \in H_*(G/PL,\ Z_2)$, represented by a Mahowald-oriented manifold M^i, and map $f : M^i \to G/PL$ with resulting surgery problem a homotopy equivalence. Then if $\underbrace{\langle x, x, \ldots, x\rangle}_{2^i}$ is defined, we have $\langle\langle x, \ldots, x\rangle, k_{(i+1)2^i-2}\rangle = 0$, where k is the Kervaire class.

(The proof is in three steps: (1) $\langle x, \ldots, x\rangle$ can be constructed in $\Omega_*(G/PL$, Mahowald orientation). Moreover, (2) each piece of the Massey product can be assumed to have corresponding surgery problem a homology equivalence. Then the surgery problem over $\langle x, \ldots, x\rangle$ is a homology equivalence, hence has Kervaire invariant 0. But (3),

$$K\langle x, \ldots, x\rangle = \langle f^*(k_*)\cdot V^2, [\langle x, \ldots, x\rangle]\rangle$$
$$= \langle f^*(k_{(i+1)2^i-2}), [\langle x, \ldots, x\rangle]\rangle = 0,$$

since $V^2 \equiv 1$ in a Mahowald-orientable manifold.)

There are only three types of homology classes which fail to admit Mahowald-orientable representatives satisfying the hypotheses of 2.3 those dual to K_{4i}, k_{4i+2}, and $Sq^1(k_{4i+2})$. For these, we require a more delicate argument. Recall that, if $\langle x_{4i+2}, k_{4i+2}\rangle = 1$, then also $\langle Q_2(x_{4i+2}), k_{8i+6}\rangle = 1$ ([10]), where Q_2 is the Araki-Kudo operation ([1]). Also, the classes y dual to K_{4i} satisfy $Q_0(y) = Q_1(y) = Q_2(y) = 0$, while those dual to k_{4i} and $Sq^1(k_{4i})$ satisfy $Q_0(y) = Q_1(y) = 0$.

We then have

Lemma 2.4. *Let y be one of the three types of classes above, and suppose*

$$\langle\underbrace{\langle y, \ldots, y\rangle}_{2^i}, k_{4j+2}\rangle = 1.$$

Then k_{8j+6} evaluates 1 on a strictly shorter Massey product.

Proof. Using the higher Mahowald orientations $M(\overline{\gamma}_i)$, we have that the theory $H_*(G/PL, M(\overline{\gamma}_i))$ admits (homology) $\cup_j$-products for $j \leq i - 1$. Thus we can apply the result of [13] and

$$(2.5)\quad Q_2\langle x, \dots, x\rangle \subset \left\langle (Q_2(x), Q_1(x), Q_0(x)) \begin{pmatrix} Q_0(x) & 0 & 0 \\ Q_1(x) & Q_0(x) & 0 \\ Q_2(x) & Q_1(x) & Q_0(x) \end{pmatrix} \dots \begin{pmatrix} Q_0(x) \\ Q_1(x) \\ Q_2(x) \end{pmatrix} \right\rangle.$$

But in all these cases, $Q_1(x) = Q_0(x) = 0$, hence this Massey product is easily seen to contain 0. Thus $Q_2\langle x, \dots, x\rangle \subset$ (indet), which runs over Massey products of length $2^j - 1$.

Remark 2.6. Actually, the hypothesis of 2.4 can be weakened to $Q_0(y) = Q_1(y) = 0$.

Now the proof of A is fairly routine. It is easily checked that the first time a differential can occur, it must hit a $(k_{4j+1})_*$. Suppose then that $d_{2^i-1}(|x|\dots|x|) = (k_{4j+2})_*$. Then $Q_2((k_{4j+2})_*)$ is $d_{2^{i-s_1}-1}(z)$, $Q_2(Q_2(k_{4j+2})_*) = d_{2^{i-s_1-s_2}-1}(z')\dots$ for $s_1, s_2\dots$ greater than 0. Finally, we will obtain $d_1(z^r) = (k_{4l+2})_*$ for some l, but this is impossible due to the fact that k_{4l+2} is primitive in $H^*(G/PL, Z_2)$ ([19]). This contradiction shows that $(k_{4j+2})_*$ is a surviving cycle for each j. Hence $E^2 = E^\infty$, and A follows.

§3. The Adams map $\mathcal{L} : B_{SO} \to G/O$

Consider the diagram (localized at 2)

$$(3.1)\qquad \begin{array}{ccccc} & & G & & \\ & & \downarrow p & & \\ B_{SO} & \xrightarrow{\mathcal{L}} & G/O & \xrightarrow{\lambda} & B_{SO} \\ & & \downarrow \Omega\tau & & \\ & & G/PL & & \end{array}$$

where $\lambda\circ\mathcal{L}$ $\psi^3 - 1$.

Lemma 3.2. <u>With coefficients</u> $Z_{(2)}$ <u>of</u> Z_2, $\mathcal{L}^*$ <u>is surjective. Indeed,</u> $\mathcal{L}^* : H^*(G/O)/_{\text{Torsion}} \to H^*(B_{SO})/_{\text{Torsion}}$ <u>is an isomorphism.</u>

We outline two proofs.

Proof. (1) Using the Bockstein spectral sequence, we find that $\{p_*(e_2 \underline{*} e_2)^{2^i}\}^*$ are integral cohomology generators. Now $p_*(e_2 \underline{*} e_2)^* = (p_*(e_1 \underline{*} e_1)^*)^2$ mod 2, and, more generally,

$$\left[\left(p_*(e_1 \underline{*} e_1)^{2^i}\right)^*\right]^2 = \left[p_*(e_2 \underline{*} e_2)^{2^i}\right]^* .$$

In dimension 2, we know that $\mathcal{L}_*(\eta) = p_*(\{\eta^2\})$ in homotopy. Hence, in cohomology, $\mathcal{L}^*((p_*(e_1 \underline{*} e_1))^*) = \omega_2$. Next, consider $Sq^1(\omega_2) = \omega_3 = \mathcal{L}^*(p_*(e_2 \underline{*} e_1)^*)$ and

$$Sq^2(Sq^1\omega_2) = \omega_2\omega_3 + \omega_5 ;$$

but $(e_1 \underline{*} e_1)^* \cup Sq^1(e_1 \underline{*} e_1)^* = (e_3 \underline{*} e_2)^* + Sq^1[(e_1 \underline{*} e_1)^2]$, while $Sq^2(Sq^1\omega_2) = (e_3 \underline{*} e_2)^* + (e_4 \underline{*} e_1)^*$. Thus using the relation

$$Sq^1\omega_4 = \omega_2\omega_3 + Sq^2Sq^1\omega_2,$$

we have $\mathcal{L}^*[p_*((e_1 \underline{*} e_1)^2)]^* = \omega_4 + \alpha$, where α is a decomposable in $\ker(Sq^1)$.

Similarly, using $Sq^4(Sq^1\omega_4) = \omega_4\omega_5 + \omega_9 + \ldots$, we obtain $\mathcal{L}^*[p_*((e_1 \underline{*} e_1)^4)]^* = \omega_8 + \alpha', \ldots, \mathcal{L}^*\left[p_*\left((e_1 \underline{*} e_1)^{2^i}\right)\right]^* = \omega_{2^{i+1}} + \alpha''$. On the other hand, as a ring over $\mathcal{A}(2)$, the ω_{2^i} generate $H^*(B_{SO})$. 3.2 now follows easily.

(2) The splitting theorem of Sullivan, $G/O = B_{SO} \times \text{Cok } J$, is proved by using the difference of the two canonical B_O-orientations on G/O to construct a map $\gamma : G/O \to B_{SO}^{\otimes}$, and $(\gamma \cdot \mathcal{L})^*$ is seen to be an isomorphism.

Lemma 3.3. $\mathcal{L}^*$ is a (rational) map of Hopf algebras.

Proof. λ is a map of H-spaces as is $\psi^3 - 1 = \lambda \cdot \mathcal{L}$. Thus the deviation of $\mathcal{L}^*$ from an H-map is contained in the torsion homology of G/O, and 3.3 follows.

Theorem 3.4. For the primitive class

$$k_{4i} \in H^*(G/PL, Z_{(2)})$$

($k_{4i} \equiv K_{4i}$ modulo decomposables), <u>we have</u>

$$(\lambda \circ \mathcal{L})^*(k_{4i}) = 2^{\alpha(i)-1} \cdot s_i$$

<u>in</u> $H^*(B_{SO}, Z_2)/_{\text{Torsion}}$, <u>where</u> s_i <u>is the primitive in</u> $H^{4i}(B_{SO}, Z_{(2)})/_{\text{Torsion}}$.

Proof. In homotopy, the generator γ_i in $\pi_{4i}(B_{SO})$ maps to the free generator β_i in $\pi_{4i}(G/O)$, and this generator maps to $q_i K_{4i}$ where K_{4i} is the generator in $\pi_{4i}(G/PL)$. From [6], q_i is the order of the subgroup $b(P_{4i}) \subset \Gamma_{4i-1}$ of homotopy spheres which bound parallelizable manifolds. This number q_i has been calculated and has the form $2^{2i-2} a_i(\text{odd})$ where $a_i = 1$ or 2. On the other hand, $\langle s_i, h(\gamma_i)\rangle = \frac{1}{2}(2i)! \cdot a_i$. Thus, since $(2i)! = 2^{2i-\alpha(i)}$, we obtain the result.

The final step in our calculation is to identify the part of $\lambda^*(k_{4i})$ contained in the torsion part of $H^*(G/O)$. (Here we are using a basis for G/O dual to the basis for $H_*(G/O)$, coming from projection of the basis for $H_*(G)$ used in [10], [14].) This we do in the next section.

§4. The image of k_{4i} and the proof of G

Consider the map $SO \to SG$. In mod 2 homology, this induces an injection $H_*(SO, Z_2) = E(e_1, e_2 \ldots) \to H_*(SG) = E(e_1 \ldots e_n \ldots) \otimes P$, where P is a polynomial algebra. In the Bockstein spectral sequence of SO, we have $E^2 = E^\infty = E(p_3, p_7, \ldots, p_{4i-1})$, where p is the primitive. However, since $H_*(SG, Q) \equiv 0$, we know that the p_{4i-1} are in the images of higher differentials in $E^*(SG)$. Indeed, there is a polynomial algebra $P(A_4, A_8, \ldots, A_{4i}, \ldots) \subset P$ so $H_*(SO) \otimes P(A_4 \ldots A_{4i} \ldots)$ is a closed sub-differential module in $E^*(SG)$. In particular, these A_{4i} in $H_*(G/O)$ generate the torsion-free part of the homology.

Definition 4.1. An element y in $H_*(X, Z_{2^i})$ is called <u>proper</u> if $2^i y = 0$ but $2^{i-1}(y) \neq 0$.

Lemma 4.2. <u>Let</u> $\mathcal{P}_i$ <u>denote the Pontrjagin squaring operator</u> $H_r(G, Z_{2^{i-1}}) \to H_{2r}(G, Z_{2^i})$. <u>Then</u> A_{4i} <u>is the mod 2 restriction of</u>

$\mathcal{P}_2^{(*)}(B_{2i})$, <u>and these</u> $\mathcal{P}_2^{(*)}(B_{2i})$, <u>together with elements</u> $\mathcal{P}_2^{(0)}(\alpha)$, <u>generate the set of proper</u> Z_4<u>-elements in</u> $H_*(SG, Z_4)$ <u>under</u> $*$ <u>and</u> $\cdot$, <u>where</u> α <u>is a proper</u> Z_2 <u>class.</u>

Similarly, we have

Lemma 4. 3. $\mathcal{P}_i \ldots [\mathcal{P}_2^*(B_{2j})]$, <u>together with iterated Pontrjagin squares</u> $\mathcal{P}_i \ldots \mathcal{P}_2(\alpha)$ <u>for</u> α <u>a proper</u> Z_2<u>-class, generate the set of proper</u> Z_{2^i}<u>-elements.</u>

Now, from the results of [3],† we see that the generators k_{4i} are primitive on proper Z_{2^i}-classes with respect to both composition and loop sum for $i > 2$. Hence the only time that $\langle k_{4i}, y\rangle$ is non-zero on a proper Z_{2^i}-class is when y is $\mathcal{P}_i \ldots [\mathcal{P}_2^*(B_{2j})]$ or $\mathcal{P}_i \ldots \mathcal{P}_2(\alpha)$. On the other hand, the fact that k_{4i} is a suspension shows that it evaluates 0 on classes of the second type. Hence it can be non-zero only on $\mathcal{P}_i \ldots [\mathcal{P}_2^*(B_{2j})]$. But these classes project into the torsion-free part of $H_*(G/O, Z_{2^i})$. This, together with 3. 4 and the remarks given in the introduction, completes the proof of Theorem G.

REFERENCES

1. S. Araki and T. Kudo. Topology of H_n-spaces and H-squaring operations. <u>Mem. Fac. Sci. Kyuoyo Univ.</u> Ser. A 10 (1956), 85-120.
2. W. Browder, A. Liulevicius and F. Peterson. Cobordism theories. <u>Ann. of Math.</u> 89 (1966), 91-101.
3. G. Brumfiel, I. Madsen and R. J. Milgram. PL characteristic classes and cobordism. <u>Ann. of Math.</u> 97 (1973), 82-159.
4. G. Brumfiel and J. Morgan. Quadratic functions, the index modulo 8, and a Z/4-Hirzebruch formula. <u>Topology</u> 12 (1973), 105-122.

† The language of [3] might lead the reader to think that the evaluation formula for (*)*(k), which is needed above, works only for Z_2-coefficients, but a little reflection will convince that it is actually valid <u>as stated</u> for all coefficients Z_{2^i}.

5. P. Conner and E. Floyd. Differentiable Periodic Maps. Ergebnesse der Mathematik 33 (Springer, 1964).

6. M. Kervaire and J. Milnor. Groups of homotopy spheres. Ann. of Math. 77 (1963), 504-537.

7. R. Kirby and L. Siebenmann. On the triangulation of manifolds and the hauptvermutung. Bull. Amer. Math. Soc. 75 (1969), 742-9.

8. S. Kochman. Thesis, Univ. of Chicago, 1970.

9. R. Lashof and M. Rothenberg. Triangulation of manifolds I, II. Bull. Amer. Math. Soc. 75 (1969), 750-7.

10. I. Madsen. Thesis, Univ. of Chicago, 1971.

11. I. Madsen and R. J. Milgram. The integral homology of B_{SG} and the 2-primary oriented cobordism ring (to appear).

12. R. J. Milgram. Iterated loop spaces. Ann. of Math. 84 (1966), 386-403.

13. R. J. Milgram. Steenrod squares and matric Massey products. Bol. Soc. Mat. Mex. 13 (1968), 32-57.

14. R. J. Milgram. The mod 2 spherical characteristic classes. Ann. of Math. 92 (1970), 238-61.

15. S. P. Novikov. Topological invariance of rational classes of Pontrjagin. Dokl. Akad. Nauk SSSR 163 (1965), 298-300.

16. D. Quillen. The Adams conjecture. Topology, 10 (1971), 67-80.

17. F. Quinn. Surgery on Poincare and normal spaces. Mimeo notes, New York Univ., 1971.

18. C. Rourke. Hauptvermutung according to Sullivan, I, II. Notes, Inst. Adv. Study, Princeton, 1968.

19. C. Rourke and D. Sullivan. On the Kervaire obstruction. Ann. of Math. 94 (1971), 397-413.

20. J. Stasheff. A classification theorem for fibre spaces. Topology, 2 (1963), 239-46.

21. D. Sullivan. The Hauptvermutung for manifolds. Bull Amer. Math. Soc. 73 (1967), 598-600.

22. D. Sullivan. Geometric Topology, Part I: Localization, Periodicity, and Galois Symmetry (M. I. T., 1970).

23. R. J. Milgram. Surgery with coefficients. Ann. of Math. (to appear).

24. I. Madsen and R. J. Milgram, Oriented PL-bordism (to appear).

Aarhus University, Aarhus, Denmark.
Stanford University, Stanford, California 94305, U. S. A.

E_∞ SPACES, GROUP COMPLETIONS, AND PERMUTATIVE CATEGORIES

J. P. MAY

In the last few years, a number of authors have developed competing theories of iterated loop spaces. Among the desirable properties of such a theory are:

(1) A recognition principle for n-fold loop spaces, $1 \leq n \leq \infty$, which applies when $n = \infty$ to such spaces at Top, BF, PL/O, etc. and to the classifying spaces of categories with appropriate structure.

(2) An approximation theorem which describes the homotopy type of $\Omega^n\Sigma^n X$, $1 \leq n \leq \infty$, in terms of iterated smash products of X and canonical spaces.

(3) A theory of homology operations on n-fold loop spaces, $1 \leq n \leq \infty$, at least sufficient to describe $H_*\Omega^n\Sigma^n X$, with all structure in sight, as a functor of H_*X.

(4) Computations and applications of the homology operations on interesting spaces to which (1) applies.

In addition, rigor and aesthetics dictate (5) and suggest (6).

(5) Complete proofs of all non-trivial technical details are to be given.

(6) Only simple and easily visualized topological constructions are to be used.

Point (5) is particularly important since several quite plausible sketched proofs of recognition principles have foundered on seemingly minor technical details. At the moment, the author's theory [17], which shall be referred to as [G], provides the only published solution to (1) and (2) which makes any claim to satisfy (5). For this reason, and because it includes the deeper cases $1 < n < \infty$ of (1) and (2) as well as machinery designed for use in (3) and (4), [G] is quite lengthy. The main purpose of the present paper is to outline and to generalize to non-connected cases the solution of (1) and (2) in the case $n = \infty$. The present proofs will not use the solution to (1) and (2) for $n < \infty$. The recognition principle will be proven in §2, after the notion of a group completion is discussed in §1. In §3 we obtain some consistency statements concerning loop spaces and classifying spaces of E_∞ spaces and discuss various approximation theorems. Finally, we demonstrate in §4 that our recognition principle applies in particular to the classifying spaces of permutative categories; the competing theories of Segal [26], Anderson [1, 2], and Tornehave [28] are designed primarily for application to such spaces. It follows that our theory can be used to construct algebraic K-theory.

In an appendix, we sharpen some of the results of [G]; these improvements are required in order to handle non-connected spaces.

I would like to emphasize that the present theory is a synthesis which incorporates many ideas borrowed from or inspired by the papers of Dyer and Lashof [11], Milgram [20], Boardman and Vogt [6, 7], Beck [5], and (in the non-connected case) Barratt, Priddy, and Quillen [3, 4, 22, 24]. My aim has been to mold these disparate lines of thought into a single coherent theory, geared towards explicit calculations in homology and homotopy and based on an absolute minimum of categorical and simplicial machinery.

The key topological applications and homological calculations based on this theory will appear in [18]. That paper will also contain a multiplicative elaboration of the theory which studies E_∞ ring spaces and applies in particular to the spaces $B\mathfrak{F}$ used in §4 to define the K-theory of a commutative topological ring.

§1. Group completions

The notion of a group completion will be central to our recognition principle for non-connected spaces. The following development is motivated by Quillen's generalization [24] of the work of Barratt and Priddy [4].

Definition 1.1. Let G be a monoid. Define the <u>translation category</u> $\tilde{G}$ of G to be the category with objects the elements of G and with morphisms from g' to g'' those elements $g \in G$ such that $g'g = g''$.

Lemma 1.2. Let G be a central submonoid of a ring R and let $i : R \to R[G^{-1}]$ denote the localization of R at G; thus $i(g)$ is invertible for $g \in G$ and i is universal with this property. Then $R[G^{-1}]$ is isomorphic as a (left and right) R-module to the limit of that functor from $\tilde{G}$ to the category of R-modules which sends each object to R and each morphism $g : g' \to g''$ to multiplication by g.

Definition 1.3. An H-space X will be said to be admissible if X is homotopy associative and if left translation by any given element of X is homotopic to right translation by the same element. A group completion $g : X \to Y$ of X is an H-map between admissible H-spaces such that Y is grouplike ($\pi_0 Y$ is a group) and the unique morphism of k-algebras

$$\bar{g}_* : H_*(X; k)[\pi_0^{-1} X] \to H_*(Y; k)$$

which extends g_* is an isomorphism for all commutative coefficient rings k.

Remark 1.4. By the following argument of Quillen [24], the condition on $\bar{g}_*$ will be satisfied if it is satisfied for $k = Z_p$ (the integers mod p) for all primes p and for $k = Q$ (the rationals). For any Abelian group A, $\pi_0 X$ acts on $H_*(X; A)$ and $H_*(X; A)[\pi_0^{-1} X]$ can be defined as the evident limit (and is a homological functor of A). Clearly, given the condition on $\bar{g}_*$ for the cited k,

$$\bar{g}_* : H_*(X; A)[\pi_0^{-1} X] \to H_*(Y; A)$$

will be an isomorphism of Abelian groups for any Q-module A, for any Z_p-module A, for $A = Z_{p^n}$ (by induction on n), and for A any torsion group (by passage to limits). Now the conclusion follows by use of the exact sequence

$$0 \to tk \to k \to k \otimes_Z Q \to t'k \to 0,$$

where tk denotes the torsion subgroup of k.

Remark 1.5. If X is a grouplike admissible H-space, then X is H-equivalent to $\pi_0 X \times X_0$, where X_0 denotes the component of the identity element [see, e.g., 18, 4.6]. Thus, by the Whitehead theorem for connected H-spaces, a group completion of a grouplike admissible H-space is a weak homotopy equivalence.

The following theorem is a special case of a result stated by Quillen [24, §9].

Theorem 1.6. Let G be a topological monoid such that G and ΩBG are admissible H-spaces. Then the natural inclusion $\zeta : G \to \Omega BG$ is a group completion.

Here B denote the standard classifying space functor (described in §3). According to Quillen, the admissibility of ΩBG need not be assumed, but I do not know how to prove the more general result. This hypothesis is clearly satisfied if G is strongly homotopy commutative, since BG is then an H-space [27, p. 251 and 269] and thus ΩBG is homotopy commutative. As I shall show in [19], Quillen's proof can be simplified under the present hypotheses.

§2. The recognition principle for E_∞ spaces

Recall from [G, 1.1 and 3.5] that an E_∞ operad $\mathcal{C}$ is a suitably compatible collection of contractible spaces $\mathcal{C}(j)$ on which the symmetric group Σ_j acts freely. Thus the orbit spaces $\mathcal{C}(j)/\Sigma_j$ are $K(\Sigma_j, 1)$'s. An action θ of $\mathcal{C}$ on a space X is a suitably compatible collection of Σ_j-equivariant maps $\theta_j : \mathcal{C}(j) \times X^j \to X$ [G, 1.2 and 1.4]. $\mathcal{C}[\mathcal{J}]$ denotes the category of $\mathcal{C}$-spaces (X, θ). An E_∞ space is a $\mathcal{C}$-space over some E_∞ operad $\mathcal{C}$. Given an E_∞ space (X, θ) and a prime p, the map $\theta_{p*} : H_*(\mathcal{C}(p) \times_{\Sigma_p} X^p) \to H_*X$ determines homology operations on H_*X. These operations will be studied and calculated on some of the spaces of primary geometric interest in [18]; see [16] for references and a partial summary.

Recall from [G, §2] that an operad $\mathcal{C}$ determines a monad (C, μ, η) such that an action θ of $\mathcal{C}$ on X is equivalent to an action

$\theta : CX \to X$ of C on X. As a space, $CX = \amalg\, \mathcal{C}(j) \times_{\Sigma_j} X^j/(\approx)$, where the equivalence relation uses base-point identifications to glue the $\mathcal{C}(j) \times_{\Sigma_j} X^j$ together (as in the proof of A. 2 below). The natural transformations $\mu : CCX \to CX$ and $\eta : X \to CX$ are given by the compatibility conditions in the definition of an operad.

A monad C can act from the right on a functor F as well as from the left on an object X [G, 9. 4]. Given such a triple (F, C, X) in the category $\mathcal{J}$ (of nice [G, p. 1] based spaces), we can construct a space $B(F, C, X)$ by forming the geometric realization ([G, 11. 1], or see the proof of A. 4 below) of the simplicial space $B_*(F, C, X)$ defined in [G, 9. 6]. The space $B_q(F, C, X)$ of q-simplices is FC^qX (where C^q is the q-fold composite); the faces are given by $FC \to F$, by $CC \to C$ applied in successive positions, and by $CX \to X$; the degeneracies are given by $1 \to C$ applied in successive positions. In an obvious sense, $B(F, C, X)$ is a functor of all three variables [G, 9. 6]. $B(F, C, X) = \amalg\, FC^qX \times \Delta_q/(\approx)$, where $\approx$ gives the appropriate face and degeneracy identifications. Thus (see [G, 9. 2 and 11. 8]) any map $\rho : Y \to FX$ determines a map $\tau(\rho) = |\tau_*(\rho)| : Y \to B(F, C, X)$ and any map $\lambda : FX \to Y$ such that $\lambda\partial_0 = \lambda\partial_1 : FCX \to Y$ determines a map $\varepsilon(\lambda) = |\varepsilon_*(\lambda)| : B(F, C, X) \to Y$. This two-sided bar construction provides all of the spaces and maps required in our theory.

We shall need to know that E_∞ spaces have group completions in order to prove that they have group completions which are infinite loop spaces.

Lemma 2. 1. Let $\mathcal{C}$ be an E_∞ operad. Then there is a functor $G : \mathcal{C}[\mathcal{J}] \to \mathcal{J}$ and a natural transformation $g : 1 \to G$ such that GX is an admissible H-space and $g : X \to GX$ is a group completion for all $\mathcal{C}$-spaces (X, θ).

Proof. As explained in [G, 3. 11], the E_∞ space (X, θ) determines an A_∞ space $(X, \theta\pi_1)$, $\pi_1 : \mathcal{C} \times \mathcal{M} \to \mathcal{C}$. Let $C \times M$ denote the monad associated to $\mathcal{C} \times \mathcal{M}$. As explained in [G, 13. 5], the A_∞ space $(X, \theta\pi_1)$ determines a topological monoid $B(M, C \times M, X)$ and maps of $\mathcal{C} \times \mathcal{M}$-spaces (in particular of H-spaces)

$$X \xleftarrow{\varepsilon(\theta\pi_1)} B(C \times M,\ C \times M,\ X) \xrightarrow{B(\pi_2,\ 1,\ 1)} B(M,\ C \times M,\ X)$$

such that $\varepsilon(\theta\pi_1)$ is a strong deformation retraction with right inverse $\tau(\eta)$. Since $\pi_2 : \mathcal{C} \times \mathcal{M} \to \mathcal{M}$ is a local Σ-equivalence (by [G, 3.7 or 3.11]), $B(\pi_2, 1, 1)$ is a homotopy equivalence by A.2 (ii) and A.4 (ii) of the appendix (which show that the connectivity hypothesis of [G, 13.5 (ii)] is unnecessary). Define

$$GX = \Omega BB(M,\ C \times M,\ X)$$

and define $g : X \to GX$ to be the composite

$$X \xrightarrow{\tau(\eta)} B(C \times M,\ C \times M,\ X) \xrightarrow{B(\pi_2,\ 1,\ 1)} B(M,\ C \times M,\ X) \xrightarrow{\zeta} GX.$$

Since X is an E_∞ space, it is certainly strongly homotopy commutative. The result follows immediately from Theorem 1.6.

Next, recall the definition, [G, 4.1], of the little cubes operads $\mathcal{C}_n$. The points of $\mathcal{C}_n(j)$ are to be thought of as j-tuples of disjoint patches on the n-sphere, and $\mathcal{C}_n(j)$ is Σ_j-equivariantly homotopy equivalent to the configuration space $F(R^n; j)$ of j-tuples of distinct points of R^n via the map which sends patches to their center points [G, 4.8]. There is an obvious natural action θ_n of $\mathcal{C}_n$ on n-fold loop spaces [G, 5.1]. (These results are due to Boardman and Vogt [7].) By use of Cohen's calculations of $H_*(F(R^n; j)/\Sigma_p)$ [10], the maps

$$(\theta_{np})_* : H_*(\mathcal{C}_n(p) \times_{\Sigma_p} (\Omega^n X)^p) \to H_*\Omega^n X$$

can be used to obtain a complete theory of homology operations on n-fold loop spaces; we recall that, for finite n and odd p, the operations of Dyer and Lashof [11] were insufficient to allow computation of $H_*\Omega^n\Sigma^n X$ and Milgram's calculation [20] of $H_*\Omega^n\Sigma^n X$ (for connected X) did not yield convenient operations.

We should point out that we are using Σ to denote the reduced suspension, whereas S was used in [G]; the purpose of this change is to emphasize by the notation that Σ^n is a functor and not a sphere. Define $\alpha_n : C_n X \to \Omega^n\Sigma^n X$ to be the composite

$$C_nX \xrightarrow{C_n\eta} C_n\Omega^n\Sigma^nX \xrightarrow{\theta_n} \Omega^n\Sigma^nX,$$

where $\eta : X \to \Omega^n\Sigma^nX$ is the standard inclusion. Our approximation theorem, proven in [G, §6 and §7], asserts that α_n is a weak homotopy equivalence if (and, by [G, 8.14], only if) X is connected. By the Whitehead theorem for connected H-spaces, this assertion in the case $n = \infty$ is also a consequence of the following result.

Theorem 2.2. $\alpha_\infty : C_\infty X \to QX = \underrightarrow{\lim}\, \Omega^n\Sigma^nX$ _is a group completion for every space_ $X \in \mathcal{J}$.

Here α_∞ is a map of $\mathcal{C}_\infty$-spaces [G, 5.2] and in particular of admissible H spaces [G, p. 4]. The homologies of $C_\infty X$ and QX with Z_p coefficients are computed in [18, 4.1 and 4.2], and it follows immediately from these calculations that the map $\overline{\alpha}_{\infty *}$ of Definition 1.3 is an isomorphism; the truth of this assertion for rational coefficients can be verified by parallel (but much simpler) computations or by appeal to the form of the E^∞-terms of the Bockstein spectral sequences of $C_\infty X$ and QX [18, 4.13]. (For connected X, H_*QX was computed by Dyer and Lashof [11].)

I am reasonably certain that $\alpha_n : C_nX \to \Omega^n\Sigma^nX$ is a group completion for all X and all n, $1 < n < \infty$, but a rigorous calculation of H_*C_nX is not yet available.

By [G, 5.2], $\alpha_n : C_n \to \Omega^n\Sigma^n$ is a morphism of monads. By [G, 9.5], it follows that the adjoint $\lambda_n : \Sigma^nC_n \to \Sigma^n$ is a C_n-functor. Therefore, if X is a $\mathcal{C}_n$-space, then $B(\Sigma^n, C_n, X)$ is defined. This space should be thought of as an n-fold de-looping of X. In particular, by [G, 13.1], $B(\Sigma^n, C_n, \Omega^nY)$ is weakly homotopy equivalent to Y if Y is n-connected and $B(\Sigma^n, C_n, C_nY)$ is homotopy equivalent to Σ^nY for any Y.

Now suppose given an arbitrary E_∞ operad $\mathcal{C}$. We wish to use the α_n to study $\mathcal{C}$-spaces. To this end, let $\mathcal{D}_n$ denote the product operad $\mathcal{C} \times \mathcal{C}_n$ [G, 3.8] and let $\psi_n : \mathcal{D}_n \to \mathcal{C}$ and $\pi_n : \mathcal{D}_n \to \mathcal{C}_n$ denote the projections. $\mathcal{D}_\infty$ is again an E_∞ operad and, if (X, θ) is a $\mathcal{C}$-space, then $(X, \theta\psi_\infty)$ is a $\mathcal{D}_\infty$-space. Thus we lose no information by studying $\mathcal{D}_\infty$-spaces instead of $\mathcal{C}$-spaces. If (X, ξ) is a $\mathcal{D}_\infty$-space, then the monad D_n acts on X via the restriction ξ_n of ξ to $D_nX \subset D_\infty X$. D_n acts on Σ^n via the composite $\lambda_n \circ \Sigma^n\pi_n$. Thus $B(\Sigma^n, D_n, X)$ is defined. We have maps of $\mathcal{D}_n$-spaces

$$X \xleftarrow{\varepsilon(\xi_n)} B(D_n, D_n, X) \xrightarrow{B(\alpha_n\pi_n, 1, 1)} B(\Omega^n\Sigma^n, D_n, X) \xrightarrow{\gamma^n} \Omega^n B(\Sigma^n, D_n, X),$$

where γ^n is defined by iteration of the obvious natural comparison $\gamma : |\Omega_*Y| \rightarrow \Omega|Y|$ for simplicial spaces Y [G, p. 115]. We are interested in the limit case, and we can define (see [G, p. 143] for the details)

$$B_iX = \lim_{\rightarrow} \Omega^j B(\Sigma^{i+j}, D_{i+j}, X).$$

Visibly $B_iX = \Omega B_{i+1}X$, and we thus obtain a functor B_∞, written $B_\infty X = \{B_iX\}$, from $\mathcal{D}_\infty[\mathcal{J}]$ to the category $\mathcal{L}_\infty$ of infinite loop sequences. Let $W : \mathcal{L}_\infty \rightarrow \mathcal{D}_\infty[\mathcal{J}]$ denote the functor given on objects $Y = \{Y_i \mid i \geq 0\} \in \mathcal{L}_\infty$ by $WY = (Y_0, \theta_\infty\pi_\infty)$, where θ_∞ is the action of $\mathcal{C}_\infty$ given in [G, 5.1]; W is to be thought of as an 'underlying E_∞ space' functor. The following recognition theorem compares the categories $\mathcal{D}_\infty[\mathcal{J}]$ and $\mathcal{L}_\infty$ by comparing WB_∞ and $B_\infty W$ to the respective identity functors. In particular, the categories of grouplike $\mathcal{D}_\infty$-spaces and of connective (Y_i is (i-1)-connected) infinite loop sequences are essentially equivalent.

Theorem 2.3. _Let (X, ξ) be a $\mathcal{D}_\infty$-space, where $\mathcal{D}_\infty = \mathcal{C} \times \mathcal{C}_\infty$ for some E_∞ operad $\mathcal{C}$. Let $\pi_\infty : \mathcal{D}_\infty \rightarrow \mathcal{C}_\infty$ be the projection. Consider the following maps of $\mathcal{D}_\infty$-spaces:_

$$X \xleftarrow{\varepsilon(\xi)} B(D_\infty, D_\infty, X) \xrightarrow{B(\alpha_\infty\pi_\infty, 1, 1)} B(Q, D_\infty, X) \xrightarrow{\gamma^\infty} WB_\infty X = B_0X.$$

(i) $\varepsilon(\xi)$ _is a strong deformation retraction with right inverse_ $\tau(\eta)$, _where_ $\eta : X \rightarrow D_\infty X$ _is given by the unit of_ D_∞.

(ii) $B(\alpha_\infty\pi_\infty, 1, 1)$ _is a group completion and is therefore a weak homotopy equivalence if_ X _is grouplike._

(iii) $\gamma^\infty : B(Q, D_\infty, X) \rightarrow B_0X$ _is a weak homotopy equivalence._

(iv) _The map_ $\iota = \gamma^\infty B(\alpha_\infty\pi_\infty, 1, 1)\tau(\eta) : X \rightarrow B_0X$ _is a group completion._

(v) B_iX is (m+i)-connected if X is m-connected.

(vi) Let $Y = \{Y_i\} \in \mathcal{L}_\infty$; there is a natural map $\omega : B_\infty WY \to Y$ in $\mathcal{L}_\infty$ such that $\omega_0 \iota = 1$ and the following diagram commutes:

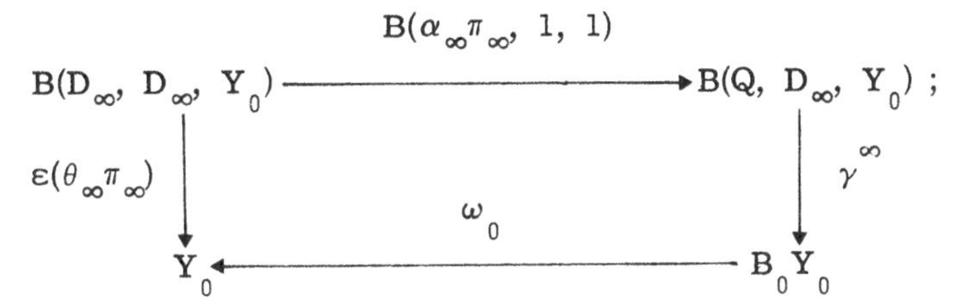

$\omega_i : B_iY_0 \to Y_i$ is a weak homotopy equivalence if Y is connective.

(vii) Let $Z \in \mathcal{J}$; then $(D_\infty Z, \mu)$ is a $\mathcal{D}_\infty$-space, where $\mu : D_\infty D_\infty Z \to D_\infty Z$ is given by the product of D_∞, and the composite

$$B_\infty D_\infty Z \xrightarrow{B_\infty \alpha_\infty \pi_\infty} B_\infty QZ \xrightarrow{\omega} Q_\infty Z = \{Q\Sigma^i Z\}$$

is a strong deformation retraction of infinite loop sequences with right inverse the adjoint $\phi_\infty(\iota\eta) : Q_\infty Z \to B_\infty D_\infty Z$ of the inclusion $\iota\eta : Z \to B_0 D_\infty Z$.

Proof. $\varepsilon(\xi)$ and $B(\alpha_\infty \pi_\infty, 1, 1)$ are realizations of maps of simplicial $\mathcal{D}_\infty$-spaces and are therefore maps of $\mathcal{D}_\infty$-spaces by [G, 12.2]. Part (i) holds before realization by [G, 9.8], hence after realization by [G, 11.10]. To prove (ii), consider the following commutative diagram:

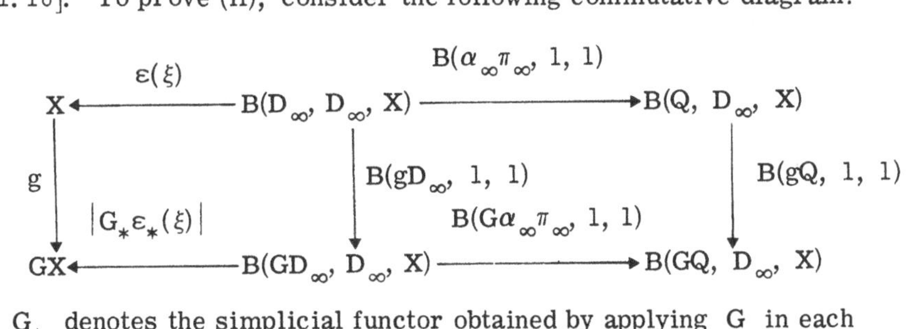

Here G_* denotes the simplicial functor obtained by applying G in each degree. $B_*(GD_\infty, D_\infty, X) = G_*B_*(D_\infty, D_\infty, X)$ by [G, 9.7], hence $|G_*\varepsilon_*(\xi)|$ makes sense; this map is a homotopy equivalence since we can apply G_* to the simplicial homotopy of [G, 9.8] and then apply [G, 11.10]. By the left-hand square, $B(gD_\infty, 1, 1)$ is a group completion since g is. By 2.2 and the fact that, by A.2 (i), $\pi_\infty : D_\infty Z \to C_\infty Z$ induces an isomorphism on homology, $\alpha_\infty \pi_\infty : D_\infty Z \to QZ$ is a group completion for any space Z. Therefore, by the very definition of a group completion,

$$G\alpha_\infty \pi_\infty : GD_\infty Z \to GQZ \quad \text{and} \quad gQ : QZ \to GQZ$$

induce isomorphisms on homology for any Z. Thus

$$B(G\alpha_\infty \pi_\infty, 1, 1) \quad \text{and} \quad B(gQ, 1, 1)$$

induce isomorphisms on homology by A.4 (i). Now (ii) follows from the right-hand square. The map γ^∞ of (iii) is obtained by passage to limits from the γ^n (see [G, p. 143]); it is a map of $\mathcal{D}_\infty$-spaces by [G, 12.4] and a weak homotopy equivalence by [G, 12.3]. Now (iv) follows from (i), (ii), and (iii), and (v) follows from [G, 11.12] (see also A.5). The ω_i in (vi) are defined by passage to loops and limits from the maps $\varepsilon\phi^n(1) : B(\Sigma^n, D_n, \Omega^n Y_n) \to Y_n$, where $\phi^n(1)$ is the evaluation; the diagram follows formally [G, p. 146 and p. 130]. Finally, (vii) follows from [G, 9.9 and 11.10], which give that $\Sigma^n Z$ is a strong deformation retract of $B(\Sigma^n, D_n, D_n Z)$, by passage to loops and limits; see [G, p. 42-43] for the adjunction ϕ_∞.

The theorem implies a uniqueness statement for the infinite loop sequence constructed from an E_∞ space.

Corollary 2.4. _Suppose given maps of $\mathcal{D}_\infty$-spaces_

$$(X, \xi) \xleftarrow{f} (X', \xi') \xrightarrow{g} (Y_0, \theta_\infty \pi_\infty)$$

such that f is a weak homotopy equivalence, g is a group completion and $Y = \{Y_i\} \in \mathcal{L}_\infty$ is connective. Then the maps

$$B_\infty X \xleftarrow{B_\infty X} B_\infty X' \xrightarrow{B_\infty g} B_\infty Y_0 \xrightarrow{\omega} Y$$

display a weak homotopy equivalence in $\mathcal{L}_\infty$ between $B_\infty X$ and Y.

Observe that there are obvious functors $\Omega^j : \mathcal{L}_\infty \to \mathcal{L}_\infty$ such that $\Omega^j_0 Y = \Omega^j Y_0$ and $\Omega^{-j}_0 Y = Y_j$ for $j \geq 0$ [G, p. 147]. The following consequence of 2.3 (vi) shows in particular that the i-th de-looping functor B_i is weakly equivalent to the i-fold iterate of B_1.

Corollary 2.5. Let (X, ξ) be a $\mathcal{D}_\infty$-space. Then the map

$$\omega_i : B_i B_j X = B_i W\Omega^{-j} B_\infty X \to \Omega_i^{-j} B_\infty X = B_{i+j} X$$

is a weak homotopy equivalence for all $i \geq 0$.

In [G, 14.4], the following further information about the homotopy type of the de-looping $B_i X$ is obtained.

Theorem 2.6. Let (X, ξ) be a $\mathcal{D}_\infty$-space. For $i > 0$, the following maps of $\mathcal{D}_\infty$-spaces are weak homotopy equivalences:

$$B(D_\infty \Sigma^i, D_\infty, X) \xrightarrow{B(\alpha_\infty \pi_\infty \Sigma^i, 1, 1)} B(Q\Sigma^i, D_\infty, X) \xrightarrow{\gamma^\infty} W\Omega^{-i} B_\infty X = B_i X.$$

As discussed in [G, p. 155], the Segal spectral sequence of $B(D_\infty \Sigma^i, D_\infty, X)$ converges to $H_* B_i X$ and has an E^2-term which, at least in principle, is a computable functor of $H_* X$.

§3. Loop spaces and classifying spaces

We shall obtain some useful consistency statements here. As an application, we shall rederive the Barratt-Quillen [3, 26] homotopy approximation to QX by relating it to the approximation we have already given in 2.3 (vii). Finally, we shall obtain a homology variant of the recognition theorem which includes Priddy's theorem [22] relating $K(\Sigma_\infty, 1)$ to QS^0.

Recall from [G, 1.5] that if $\mathcal{C}$ is any operad and if (X, θ) is a $\mathcal{C}$-space, then ΩX is again a $\mathcal{C}$-space with action defined pointwise and denoted by $\Omega\theta$. By iteration, we have functors $\Omega^i : \mathcal{C}[\mathcal{J}] \to \mathcal{C}[\mathcal{J}]$ for $i > 0$.

Theorem 3.1. Let $\mathcal{C}$ be an E_∞ operad and let $\mathcal{D}_\infty = \mathcal{C} \times \mathcal{C}_\infty$ with projections $\psi_\infty : \mathcal{D}_\infty \to \mathcal{C}$ and $\pi_\infty : \mathcal{D}_\infty \to \mathcal{C}$. Let (X, θ) be a $\mathcal{C}$-space. Then, for $i > 0$, there is a $\mathcal{D}_\infty$-space $Y_i X$ and there are maps of $\mathcal{D}_\infty$-spaces

$$(X, \theta\psi_\infty) \xleftarrow{\varepsilon} Y_i X \xrightarrow{\delta} (B_i\Omega^i X, \theta_\infty\pi_\infty) = W\Omega^{-i}B_\infty\Omega^i X$$

such that $\Omega^i\varepsilon$ and δ are weak homotopy equivalences. Therefore, if X is (i-1)-connected, then the infinite loop sequences $B_\infty X$ and $\Omega^{-i}B_\infty\Omega^i X$ are weakly homotopy equivalent.

Proof. The second statement will follow from the first by use of 2. 3 (vi). The basic point is that when taking limits of loop spaces (via $\Omega^n X \to \Omega^{n+1}\Sigma X$) the new coordinate is the last coordinate, whereas when forming loop spaces the new coordinate is the first coordinate. Since X is a $\mathcal{D}_\infty$-space via $\theta\psi_\infty$ and $\Omega^i X$ is a $\mathcal{D}_\infty$-space via $\Omega^i(\theta\psi_\infty) = (\Omega^i\theta)\psi_\infty$ and since the suspensions $\mathcal{D}_n \to \mathcal{D}_{n+1}$ involve only the little cubes coordinates, it is plausible that our categorical constructions can be suspended so as to free the first i coordinates. The detailed constructions occupy much of [G, §14] and $Y_i X$, ε, and δ are specified in the statement of [G, 14. 9]. We need only indicate here how the connectivity hypothesis of [G, 14. 9] can be improved. In the proof of [G, 14. 7], we must show that the map $B(\delta_{i\infty}\tau'_{i\infty}, 1, 1)$ in the bottom diagram of [G, p. 150] is a weak homotopy equivalence (which will imply that $\Omega^i\varepsilon$ of the present statement is a weak homotopy equivalence). To prove this, consider the following commutative diagram ($\xi_i = \Omega^i\theta\cdot\psi_\infty$):

$$\begin{array}{ccccc}
\Omega^i X & \xleftarrow{\varepsilon(\xi_i)} & B(D_\infty, D_\infty, \Omega^i X & \xrightarrow{B(\delta_{i\infty}\tau'_{i\infty}, 1, 1)} & B(\Omega^i D'_\infty\Sigma^i, D_\infty, \Omega^i X) \\
\downarrow g & & \downarrow B(gD_\infty, 1, 1) & & \downarrow B(g\Omega^i D'_\infty\Sigma^i, 1, 1) \\
G\Omega^i X & \xleftarrow{|G_*\varepsilon_*(\xi_i)|} & B(GD_\infty, D_\infty, \Omega^i X) & \xrightarrow{B(G\delta_{i\infty}\tau'_{i\infty}, 1, 1)} & B(G\Omega^i D'_\infty\Sigma^i, D_\infty, \Omega^i X)
\end{array}$$

Here $\delta_{i\infty}\tau'_{i\infty} : D_\infty Z \to \Omega^i D'_\infty \Sigma^i Z$ is a group completion for any Z by [G, 14. 2], 2. 2, and A. 2 (applied to the local Σ-equivalence $\tau'_{i\infty}$). By arguments precisely analogous to those used to prove 2. 3 (ii), $B(\delta_{i\infty}\tau'_{i\infty}, 1, 1)$ is a group completion and thus, since $\Omega^i X$ is grouplike, a weak homotopy equivalence. Again, connectivity was assumed in [G, 14. 8] because of references to [G, 3. 4 and 11. 13], and we can now refer to A. 2 and A. 4 instead. With these changes, the result follows as in the proof of [G, 14. 9].

Remark 3.3. If $Y \in \mathcal{L}_\infty$ and $i > 0$, then $\Omega^i WY$ and $W\Omega^i Y$ are $\Omega^i Y_0$ together with two different actions $D_\infty \Omega^i Y_0 \to \Omega^i Y_0$. By [G, 14.10] and A.2, these action maps are homotopic.

To relate our de-loopings to the standard classifying space functor, we require some recollections from [G, §10]. If $\mathcal{U}$ is any category with finite products, then the notions of a monoid G in $\mathcal{U}$ and of right and left G-objects Y and X in $\mathcal{U}$ are defined. There is an obvious category $\mathcal{A}\mathcal{U}$ with objects such triples (Y, G, X) and there is a two-sided simplicial bar construction $B_* : \mathcal{A}\mathcal{U} \to \mathcal{S}\mathcal{U}$. When $\mathcal{U}$ is the category of (nice, unbased) spaces, we can compose B_* with geometric realization to obtain a functor $B : \mathcal{A}\mathcal{U} \to \mathcal{U}$. Indeed, this construction is just another application of that used in the previous section. We shall use B(Y, G, X) to study the classification of various types of fibrations in [19]. Let $\delta : G \to *$ be the unique map onto the one-point G-space $*$ and define

$$p = B(1, 1, \delta) : EG = B(*, G, G) \to B(*, G, *) = BG.$$

BG is the standard classifying space of the monoid G, EG is a contractible right G-space, and p is a principal quasi G-fibration if G is grouplike (or G-bundle if G is a group).

Now, until otherwise specified, let $\mathcal{C}$ be any operad. By [G, 1.6 and 1.7], $\mathcal{C}[\mathcal{T}]$ has finite products. By [G, 12.2], the geometric realization of a simplicial $\mathcal{C}$-space is a $\mathcal{C}$-space. Thus $B? = |B_* ?|$ defines a functor $\mathcal{A}\mathcal{C}[\mathcal{T}] \to \mathcal{C}[\mathcal{T}]$. An object (Y, G, X) of $\mathcal{A}\mathcal{C}[\mathcal{T}]$ consists of a topological monoid G and right and left G-spaces Y and X such that Y, G, and X are $\mathcal{C}$-spaces and the product and unit of G and the actions of G on Y and on X are maps of $\mathcal{C}$-spaces. For clarity, write (G, θ, ϕ) for a monoid in $\mathcal{C}[\mathcal{T}]$, where θ is the action of $\mathcal{C}$ and ϕ is the monoid product. The unit condition ensures that the base-point (for θ) coincides with the identity element e (for ϕ). The product ϕ need not be (and in practice is not) $\theta_2(c)$ for any element $c \in \mathcal{C}(2)$, but we have the following observation.

Lemma 3.4. _Let_ (G, θ, ϕ) _be a monoid in_ $\mathcal{C}[\mathcal{T}]$. _If_ $\mathcal{C}(1)$ _is connected, then_ ϕ _is homotopic to_ $\theta_2(c)$ _for any_ $c \in \mathcal{C}(2)$.

Proof. Write $\phi(g, g') = gg'$ and $\theta_2(c)(g, g') = g \wedge g'$. Since ϕ is a map of $\mathcal{C}$-spaces, $(g_1g_2) \wedge (g_1'g_2') = (g_1 \wedge g_1')(g_2 \wedge g_2')$ and therefore $g_1 \wedge g_2' = (g_1 \wedge e)(e \wedge g_2')$. Since e is a two-sided homotopy identity for $\wedge$, by [G, p. 4], the conclusion follows.

The following result asserts the existence and essential uniqueness of classifying spaces in $\mathcal{C}[\mathcal{T}]$ for monoids in $\mathcal{C}[\mathcal{T}]$.

Proposition 3.5. *Let (G, θ, ϕ) be a monoid in $\mathcal{C}[\mathcal{T}]$. Then BG and EG admit actions $B\theta$ and $E\theta$ of $\mathcal{C}$ such that EG is a right G-space in $\mathcal{C}[\mathcal{T}]$ and $p : EG \to BG$ is a map of $\mathcal{C}$-spaces. If G is grouplike and if $p' : E' \to B'$ is a map of $\mathcal{C}$-spaces and a principal quasi G-fibration such that E' is contractible and is a right G-space in $\mathcal{C}[\mathcal{T}]$, then B' is weakly homotopy equivalent as a $\mathcal{C}$-space to BG.*

Proof. Consider the following commutative diagram:

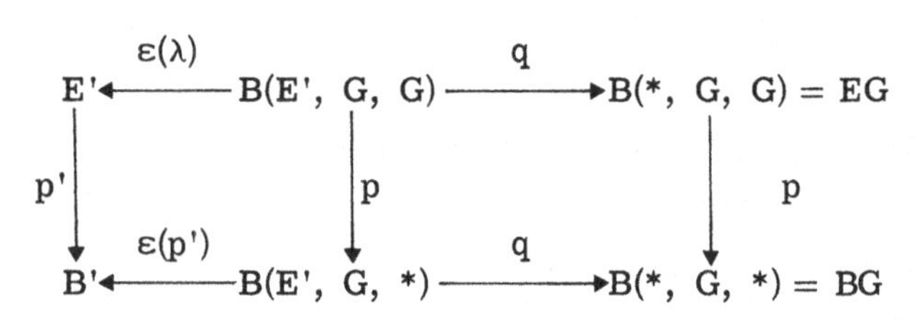

Here $\lambda : E' \times G \to E'$ is the given right action, $\varepsilon(?) = |\varepsilon_*(?)|$ with ε_* as in [G, 9.2], $p = B(1, 1, \delta)$, and $q = B(\delta, 1, 1)$. All maps are realizations of maps of simplicial $\mathcal{C}$-spaces and $\varepsilon(\lambda)$ and q are realizations of maps of simplicial right G-spaces in $\mathcal{C}[\mathcal{T}]$. Since p' and (see [19]) both maps p are principal quasi G-fibrations with contractible total spaces, $\varepsilon(p')$ and the bottom map q are weak homotopy equivalences.

The comparison of G to ΩBG can also be carried out in $\mathcal{C}[\mathcal{T}]$.

Proposition 3.6. *Let (G, θ, ϕ) be a monoid in $\mathcal{C}[\mathcal{T}]$. Then the natural inclusion $\zeta : G \to \Omega BG$ is a map of $\mathcal{C}$-spaces.*

Proof. $\zeta(g)(t) = |[g], (t, 1\text{-}t)|$ for $g \in G$ and $t \in I$, where $[g]$ is g regarded as a 1-simplex of B_*G and $(t, 1\text{-}t) \in \Delta_1$. $B\theta$ is the composite

$$CBG = C|B_*G| \xrightarrow{\nu^{-1}} |C_*B_*G| \xrightarrow{|B_*\theta|} |B_*G| = BG,$$

where $B_*\theta$ is the simplicial action specified on q-simplices by $B_q\theta = \theta^q$ and where ν is the homeomorphism of [G, 12.2]. Therefore, if $c \in \mathcal{C}(j)$, $g_i \in g$, and $t \in I$, then

$$\begin{aligned}(\Omega B\theta)_j(c, \zeta(g_1), \ldots, \zeta(g_j))(t) &= (B\theta)_j(c, \zeta(g_1)(t), \ldots, \zeta(g_j)(t)) \\ &= (B\theta)_j(c, |[g_1], (t, 1-t)|, \ldots, |[g_j], (t, 1-t)|) \\ &= |B_*\theta|\,|[c; [g_1], \ldots, [g_j]], (t, 1-t)| \\ &= |[\theta_j(c, g_1, \ldots, g_j)], (t, 1-t)| \\ &= \zeta\theta_j(c, g_1, \ldots, g_j)(t)\end{aligned}$$

(Here $[c; [g_1], \ldots, [g_j]]$ is an element of $CB_1G = CG$.)

Returning to the context of E_∞ spaces, we can now compare our de-looping B_1 to the classifying space functor B.

Theorem 3.7. Let $\mathcal{C}$ be an E_∞ operad and let $\mathcal{D}_\infty = \mathcal{C} \times \mathcal{C}_\infty$ with projection $\psi_\infty : \mathcal{D}_\infty \to \mathcal{C}$. Let (G, θ, ϕ) be a monoid in $\mathcal{C}[\mathcal{T}]$. Then $(\Omega BG, \Omega B\theta \circ \psi_\infty)$ is weakly homotopy equivalent as a $\mathcal{D}_\infty$-space to $WB_\infty G$ and $(BG, B\theta \circ \psi_\infty)$ is weakly homotopy equivalent as a $\mathcal{D}_\infty$-space to $W\Omega^{-1}B_\infty G$ (which has underlying space B_1G). Therefore the infinite loop sequences $B_\infty\Omega BG$, $B_\infty G$, and $\Omega B_\infty BG$ are all weakly homotopy equivalent.

Proof. Since $\zeta : G \to \Omega BG$ is a map of $\mathcal{D}_\infty$-spaces and a group completion, $B_\infty\zeta : B_\infty G \to B_\infty\Omega BG$ is defined and is a weak homotopy equivalence of infinite loop sequences. This obviously implies that $B_i\zeta : W\Omega^{-i}B_\infty G \to W\Omega^{-i}B_\infty\Omega BG$ is a weak homotopy equivalence of $\mathcal{D}_\infty$-spaces for $i \geq 0$. When $i = 0$, $WB_\infty\Omega BG$ is weakly homotopy equivalent as a $\mathcal{D}_\infty$-space to ΩBG by 2.3. When $i = 1$, $W\Omega^{-1}B_\infty\Omega BG$ is weakly homotopy equivalent as a $\mathcal{D}_\infty$-space to BG by 3.2. The last statement follows by use of 2.3 (vi).

Barratt [3] and Quillen [unpublished; see Segal [26]] have given approximations to QX. The key fact behind any such result is 2.2: any natural group completion of CX, where $\mathcal{C}$ is an E_∞ operad, should approximate QX. In this sense, 2.3 (vii) is another such approximation theorem and so is [G, 6.4], which asserts that $\Omega C\Sigma X$ approximates QX. Barratt and Quillen focus attention (implicitly, see [G, 6.5]) on one particular E_∞ operad, namely $\mathcal{D} = \mathcal{D}\mathcal{M}$, as defined in [G, p. 161]. This is a 'minimal' E_∞ operad in that $\mathcal{D}(1)$ is a point; $\mathcal{D}(j)$ is the normalized version of Milnor's universal bundle for Σ_j (see 4.7 and 4.8). Let (D, μ, η) be the monad associated to $\mathcal{D}$. As shown in [G, p. 161], DX is a topological monoid; denote its product by $\oplus$.

Theorem 3.7. For all $X \in \mathcal{T}$, $(DX, \mu, \oplus)$ is a monoid in $\mathcal{D}[\mathcal{T}]$ and ΩBDX is weakly homotopy equivalent as a $\mathcal{D}_\infty$-space to QX (that is, to $WQ_\infty X$), where $\mathcal{D}_\infty = \mathcal{D} \times \mathcal{C}_\infty$.

Proof. Let $c \in \mathcal{D}(j)$ and let $[d_i; y_i], [e_i, z_i] \in DX$, $1 \leq i \leq j$. By [G; 1.7, 2.4 (iii), and p. 161], in order to show that $\oplus$ is a map of $\mathcal{D}$-spaces we must verify the formula

$$[\gamma(c; d_1, \ldots, d_j) \oplus \gamma(c; e_1, \ldots, e_j); y_1, \ldots, y_j, z_1, \ldots, z_j]$$
$$= [\gamma(c; d_1 \oplus e_1, \ldots, d_j \oplus e_j); y_1, z_1, \ldots, y_j, z_j].$$

By [G, 15.1], γ is obtained by applying the product-preserving functor $|D_*?|$ (see [G, 10.2]) to γ for the operad $\mathcal{M}$ of [G, 3.1]. On the level of symmetric groups (that is, in $\mathcal{M}$), if $\tau \oplus \tau'$ denotes the permutation of successive blocks of letters determined by τ and τ', then

$$\gamma(\sigma; \tau_1, \ldots, \tau_j) = \tau_{\sigma^{-1}(1)} \oplus \ldots \oplus \tau_{\sigma^{-1}(j)},$$

hence

$$\gamma(\sigma; \tau_1, \ldots, \tau_j) \oplus \gamma(\sigma; \mu_1, \ldots, \mu_j) = \gamma(\sigma; \tau_1 \oplus \mu_1, \ldots, \tau_j \oplus \mu_j)\nu$$

for a certain permutation ν (which shuffles blocks of letters). It follows (by a diagram chase) that the same relation holds in $\mathcal{D}$, and the desired equality results in view of the equivariance identifications used to define DX. By the previous theorem, ΩBDX is weakly homotopy equivalent as a $\mathcal{D}_\infty$-space to $WB_\infty DX$. The projection $\psi_\infty : D_\infty X \to DX$ is a map of $\mathcal{D}_\infty$-spaces and induces an isomorphism on homology by A. 2 (i). By 2. 3, $B_0\psi_\infty : WB_\infty D_\infty X \to WB_\infty DX$ induces an isomorphism on homology and is therefore a weak homotopy equivalence of $\mathcal{D}_\infty$-spaces. Now the conclusion follows from 2. 3 (vii).

The most striking instance of the theorem is the case $X = S^0$, when, by [G, 8. 11], $DS^0 = \amalg \mathcal{D}(j)/\Sigma_j$ and the assertion is that QS^0 is homotopy equivalent to the group completion $\Omega B \amalg K(\Sigma_j, 1)$. We shall obtain a related homology approximation to QX in [18, §5]; when $X = S^0$, this result will reduce to Priddy's theorem [22] which states that $K(\Sigma_\infty, 1) \times Z$ is homologically equivalent to QS^0. We here give another instance of the same type of homology approximation which also contains Priddy's theorem.

Construction 3. 8. Let G be the free monoid on one generator g and let $\mathcal{C}$ be any operad. Consider the category an object of which is a $\mathcal{C}$-space (X, θ) together with an inclusion of monoids $G \subset \pi_0 X$ and a chosen base-point $a \in g$; morphisms are to preserve all of these data. Construct a functor from this category to spaces as follows. Fix an element $c \in \mathcal{C}(2)$. Define $\rho(a) : X \to X$ by $\rho(a)(x) = \theta_2(x)(x, a)$; thus, $\rho(a)$ is right translation by a. Assume (for simplicity) or arrange (by use of mapping cylinders) that $\rho(a) : X_n \to X_{n+1}$ is a cofibration, where X_n denotes the component g^n, $n \geq 0$. Then define $\overline{X}$ to be the limit of the X_n under the maps $\rho(a)$; clearly morphisms f in our domain category determine maps $\overline{f}$ by passage to limits.

Proposition 3. 9. _Let_ $\mathcal{D}_\infty = \mathcal{C} \times \mathcal{C}_\infty$ _where_ $\mathcal{C}$ _is an_ E_∞ _operad such that_ $\mathcal{C}(j)$, $j \geq 1$, _has the_ Σ_j_-equivariant homotopy type of a_ CW _free_ Σ_j_-complex. Let_ (X, ξ) _be a_ $\mathcal{D}_\infty$_-space such that_ X _has the homotopy type of a_ CW_-complex and_ $\pi_0 X$ _is a free monoid on one generator_ g. _Let_ $(B_0 X)_0$ _denote the component of the base-point of_ $B_0 X$. _Then there is a map_ $\overline{\iota} : \overline{X} \to (B_0 X)_0$ _such that_ $\overline{\iota}_* : H_*(\overline{X}; k) \to H_*((B_0 X)_0; k)$ _is an isomorphism of algebras for all commutative rings_ k.

Proof. By 3.8 applied to the operad $\mathcal{D}_\infty$, we have spaces and maps

$$\overline{X} \xleftarrow{\overline{\varepsilon}(\xi)} \overline{B}(D_\infty, D_\infty, X) \xrightarrow{\overline{B}(\alpha_\infty \pi_\infty, 1, 1)} \overline{B}(Q, D_\infty, X) \xrightarrow{\overline{\gamma}^\infty} \overline{B}_0 X \, ;$$

the relevant inclusions of $G = \pi_0 X$ are evident. By A.3, A.6, and Milnor's theorem [21], all spaces in sight have the homotopy type of CW-complexes. $\overline{\varepsilon}(\xi)$ and the natural inclusion of $(B_0 X)_0$ in $\overline{B}_0 X$ are homotopy equivalences (since $\varepsilon(\xi)$ and each $\rho(a) : (B_0 X)_n \to (B_0 X)_{n+1}$ are homotopy equivalences); choose homotopy inverses $\overline{\tau}(\eta)$ and λ and define

$$\overline{\iota} = \lambda \circ \overline{\gamma}^\infty \circ \overline{B}(\alpha_\infty \pi_\infty, 1, 1) \circ \overline{\tau}(\eta) \, .$$

$H_*(\overline{X}; k)$ is a well-defined algebra since $\pi_0 X$ is central in the associative algebra $H_*(X; k)$, and $\overline{\iota}_*(x) = x \circ g^{-n}$ for $x \in H_*(X_n; k)$ since the restriction of λ to $(B_0 X)_n$ is homotopic to right translation by any point in the component $g^{-n} \in \pi_0 B_0 X$. For any $\mathcal{D}_\infty$-space X, 1.2, 1.5, and 2.3 imply that

$$H_*(B_0 X; k) = k\pi_0 B_0 X \otimes H_*((B_0 X)_0; k) \quad \text{and} \quad H_*((B_0 X)_0; k) \cong \varinjlim H_*(X_a; k)$$

where $k\pi_0 B_0 X$ is the group ring and the limit is taken over the translation functor from $\widetilde{\pi_0 X}$ to k-modules which sends the object $a \in \widetilde{\pi_0 X}$ to $H_*(X_a; k)$ and the morphism $b : a \to a'$ to multiplication by b. In our case, the isomorphism is realized as $\lambda_*^{-1} : H_*((B_0 X)_0; k) \to H_*(\overline{B}_0 X; k)$, and the desired conclusion follows from the definition of $\overline{X}$.

Priddy's theorem is obtained by taking $X = D_\infty S^0$ (or $X = CS^0$), since then $\overline{X}$ is a $K(\Sigma_\infty, 1)$ and $B_0 X$ is homotopy equivalent to QS^0.

§4. Symmetric monoidal and permutative categories

Our goal here is to demonstrate that our theory associates infinite loop spaces with good properties to categories of the specified types by observing that the classifying space of a permutative category is naturally an E_∞ space. This observation (and it is no more than that: the proof is a triviality) has been known to Stasheff and myself for some time; it only acquires usefulness with the present extension of my theory to non-connected spaces. We shall also indicate how to use our infinite loop spaces to define algebraic K-theory and shall compute our K^0R, $K^{-1}R$, and $K^{-2}R$ for a topological ring R.

Recall that a topological category $\mathcal{A}$ is a small category in which the set $\mathcal{O}\mathcal{A}$ of objects of $\mathcal{A}$ and the set $\mathcal{M}\mathcal{A}$ of morphisms of $\mathcal{A}$ are (compactly generated Hausdorff) spaces and the four structural functions

$$\text{Source } S: \mathcal{M}\mathcal{A} \to \mathcal{O}\mathcal{A}, \qquad \text{Target } T: \mathcal{M}\mathcal{A} \to \mathcal{O}\mathcal{A}$$
$$\text{Identity } I: \mathcal{O}\mathcal{A} \to \mathcal{M}\mathcal{A}, \qquad \text{Composition } C: \mathcal{M}\mathcal{A} \times_{\mathcal{O}\mathcal{A}} \mathcal{M}\mathcal{A} \to \mathcal{M}\mathcal{A}$$

are continuous, where $\mathcal{M}\mathcal{A} \times_{\mathcal{O}\mathcal{A}} \mathcal{M}\mathcal{A} = \{(g, f) | f, g \in \mathcal{M}\mathcal{A}, Sg = Tf\}$. Henceforward, in all definitions, theorems, etc., all given categories are tacitly assumed to be topological and all given functors and natural transformations are tacitly assumed to be continuous; all constructed gadgets must be proven to be consistent with the topology. Of course, if no topology is in sight, we can always impose the discrete topology.

Monoidal, strict monoidal, and symmetric monoidal categories are defined in [14, VII §1 and §7]. Provided that the collection of isomorphism classes of objects forms a set, we can replace a given large, hence non-topological, (symmetric) monoidal category by an equivalent small (symmetric) monoidal category simply by choosing any skeleton [14, p. 91].

Definition 4.1. <u>A permutative category</u> $(\mathcal{A}, \square, *, c)$ is a symmetric strict monoidal category. In detail, $\square : \mathcal{A} \times \mathcal{A} \to \mathcal{A}$ is an associative bifunctor, $* \in \mathcal{O}\mathcal{A}$ is a two-sided identity for $\square$, and $c : \square \to \square\tau$ is a natural transformation (where $\tau : \mathcal{A} \times \mathcal{A} \to \mathcal{A} \times \mathcal{A}$ is the transposition) such that $c^2 = 1$, $c(A\square *) = IA$ for $A \in \mathcal{O}\mathcal{A}$, and the following diagram is commutative for $A, B, C \in \mathcal{O}\mathcal{A}$:

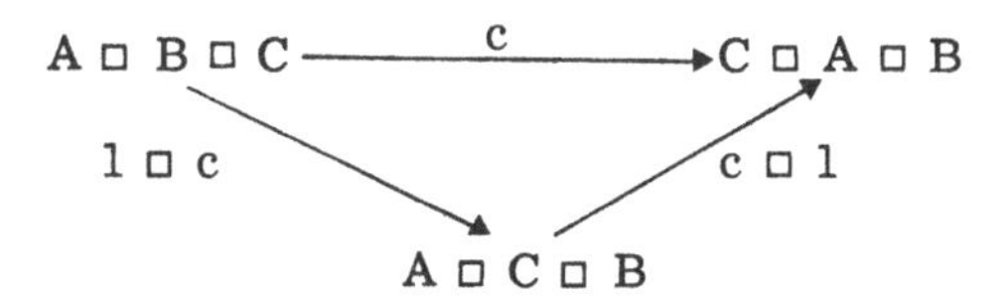

By a trivial special case of MacLane's coherence theorem [13, 5.1], all diagrams built up from $\Box$, *, and c are commutative.

Following MacLane [14], we are using the neutral symbol $\Box$ rather than $\otimes$ or $\oplus$, which are distractingly suggestive of special cases. Symmetric monoidal categories are ubiquitous but permutative categories are seldom found in nature. This is no limitation in view of the following result, which is due to Isbell [12].

Proposition 4.2. Let $\mathcal{A}$ be a monoidal category. Then there is a naturally equivalent strict monoidal category $\mathcal{B}$; if $\mathcal{A}$ is symmetric, then $\mathcal{B}$ is permutative.

Proof. Define $\mathcal{O}\mathcal{B} = M\mathcal{O}\mathcal{A}$ as a topological monoid where $M\mathcal{O}\mathcal{A}$, the James construction on $\mathcal{O}\mathcal{A}$, is the free topological monoid generated by $\mathcal{O}\mathcal{A}$ subject to the single relation $* = e$ [G, 3.2]. Let $\eta : \mathcal{O}\mathcal{A} \to \mathcal{O}\mathcal{B}$ be the standard inclusion. Write objects of $\mathcal{B}$ by juxtaposition of objects of $\mathcal{A}$ and define $\pi : \mathcal{O}\mathcal{B} \to \mathcal{O}\mathcal{A}$ by

$$\pi(A_1 \dots A_n) = A_1 \Box (A_2 \Box (A_3 \Box \dots (A_{n-1} \Box A_n) \dots)), \quad A_i \in \mathcal{O}\mathcal{A}.$$

Define $\mathcal{M}\mathcal{B}$ by $\mathcal{B}(B, B') = \{B\} \times \mathcal{A}(\pi B, \pi B') \times \{B'\}$. The singleton sets are required for disjointness of hom sets and they determine S and T for $\mathcal{B}$; I and C for $\mathcal{B}$ are induced from I and C for $\mathcal{A}$. $\mathcal{M}\mathcal{B}$ is topologized as a subspace of $\mathcal{O}\mathcal{B} \times \mathcal{M}\mathcal{A} \times \mathcal{O}\mathcal{B}$, and the four functions are clearly continuous. $\Box : \mathcal{M}\mathcal{B} \times \mathcal{M}\mathcal{B} \to \mathcal{M}\mathcal{B}$ and, if $\mathcal{A}$ is symmetric, the symmetry c of $\mathcal{B}$ are determined by the following morphisms of $\mathcal{A}$:

$$\pi(B \Box C) \xrightarrow{\cong} \pi B \Box \pi C \xrightarrow{f \Box g} \pi B' \Box \pi C' \xrightarrow{\cong} \pi(B' \Box C')$$

for morphisms (B, f, B') and (C, g, C') of $\mathcal{B}$ and

$$\pi(B \Box C) \xrightarrow{\cong} \pi B \Box \pi C \xrightarrow{\quad c \quad} \pi C \Box \pi B \xrightarrow{\cong} \pi(C \Box B)$$

for objects B and C of $\mathcal{B}$; the unlabelled isomorphisms are uniquely determined by the monoidal structure of $\mathcal{A}$. Define $\eta : \mathcal{M}\mathcal{A} \to \mathcal{M}\mathcal{B}$ by $\eta(f) = (A, f, A')$ for $f : A \to A'$ and define $\pi : \mathcal{M}\mathcal{B} \to \mathcal{M}\mathcal{A}$ by

$\pi(B, g, B') = g$ for $g : \pi B \to \pi B'$. Then η and π are functors such that $\pi\eta = 1$, and the morphisms $(B, I\pi B, \eta\pi B)$ of $\mathcal{B}$ define a natural isomorphism between 1 and $\eta\pi$.

Let $(\mathcal{A}, \square, *, c)$ be a fixed permutative category. For $\sigma \in \Sigma_j$, c determines a natural transformation $c_\sigma : \square_j \to \square_j \sigma$, where Σ_j acts on the left of $\mathcal{A}^j$ by permutation of coordinates and $\square_j : \mathcal{A}^j \to \mathcal{A}$ denotes the iterate of $\square$; by coherence, $c_\sigma c_\tau = c_{\sigma\tau}$. We have the following three lemmas, the first of which is an observation due to Anderson [2]; the other two are direct consequences of coherence: we need only observe that the assertions make sense on objects. Recall the definition, 1.1, of the translation category $\tilde{G}$ of a monoid G. Observe that G acts on the right of $\tilde{G}$ via the product of G. Observe too that if G is a group, then there is a unique morphism $g \to g'$ for $g, g' \in G$ and a functor with range $\tilde{G}$ is therefore uniquely determined by its object function.

Lemma 4.3. _For_ $j \geq 0$, _there is a_ Σ_j_-equivariant functor_

$$c_j : \tilde{\Sigma}_j \times \mathcal{A}^j \to \mathcal{A}$$

defined, on objects and morphisms respectively, by

$$c_j(\sigma, A_1, \ldots, A_j) = A_{\sigma^{-1}(1)} \square \ldots \square A_{\sigma^{-1}(j)}$$

and

$$c_j(\sigma \to \tau, f_1, \ldots, f_j) = c_{\tau\sigma^{-1}} \circ (f_{\sigma^{-1}(1)} \square \ldots \square f_{\sigma^{-1}(j)}).$$

(If $j = 0$, $\tilde{\Sigma}_0 \times \mathcal{A}^0$ is the unit category, with one object and one morphism, and c_0 is the functor determined by $* \in \mathcal{O}\mathcal{A}$.)

Lemma 4.4. _The following diagram is commutative for all_ $j \geq 0$, $k \geq 0$, _and_ $j_i \geq 0$ _with_ $j_1 + \ldots + j_k = j$:

$$
\begin{array}{ccccc}
\tilde{\Sigma}_k \times \tilde{\Sigma}_{j_1} \times \dots \times \tilde{\Sigma}_{j_k} \times \mathcal{A} & \xrightarrow{\tilde{\gamma} \times 1} & \tilde{\Sigma}_j \times \mathcal{A}^j & & \\
 & & & \searrow^{c_j} & \\
\Big\downarrow {\scriptstyle 1 \times \mu} & & & & \mathcal{A} \\
 & & & \nearrow_{c_k} & \\
\tilde{\Sigma}_k \times \tilde{\Sigma}_{j_1} \times \mathcal{A}^{j_1} \times \dots \times \tilde{\Sigma}_{j_k} \times \mathcal{A}^{j_k} & \xrightarrow{1 \times c_{j_1} \times \dots \times c_{j_k}} & \tilde{\Sigma}_k \times \mathcal{A}^k & &
\end{array}
$$

where μ is the evident shuffle isomorphism and the functors $\tilde{\gamma} : \tilde{\Sigma}_k \times \tilde{\Sigma}_{j_1} \times \dots \times \tilde{\Sigma}_{j_k} \to \tilde{\Sigma}_j$ are defined on objects by

$$\tilde{\gamma}(\sigma; \tau_1, \dots, \tau_k) = \tau_{\sigma^{-1}(1)} \oplus \dots \oplus \tau_{\sigma^{-1}(k)}.$$

Lemma 4.5. For $j \geq 0$, c_ν determines a natural isomorphism between the two composites in the following diagram:

$$
\begin{array}{ccccc}
\tilde{\Sigma}_j \times (\mathcal{A} \times \mathcal{A})^j & \xrightarrow{\Delta \times \nu} & \tilde{\Sigma}_j \times \tilde{\Sigma}_j \times \mathcal{A}^j \times \mathcal{A}^j & \xrightarrow{1 \times \tau \times 1} & \tilde{\Sigma}_j \times \mathcal{A}^j \times \tilde{\Sigma}_j \times \mathcal{A}^j \\
\Big\downarrow {\scriptstyle 1 \times \Box^j} & & & & \Big\downarrow {\scriptstyle c_j \times c_j} \\
\tilde{\Sigma}_j \times \mathcal{A}^j & \xrightarrow{c_j} & \mathcal{A} & \xleftarrow{\Box} & \mathcal{A} \times \mathcal{A}
\end{array}
$$

Here $\nu \in \Sigma_{2j}$ determines the evident shuffle isomorphism, τ is the transposition, and Δ is the diagonal functor.

Now recall the following definition, due to Segal [25].

Definition 4.6. Let $\mathcal{A}$ be a category. The nerve, or morphism complex, $B_*\mathcal{A}$ of $\mathcal{A}$ is the simplicial space specified as follows:

$$B_0\mathcal{A} = \mathcal{O}\mathcal{A}, \quad B_1\mathcal{A} = \mathcal{M}\mathcal{A}, \quad \text{and if } q > 1$$

$$B_q\mathcal{A} = \mathcal{M}\mathcal{A} \times_{\mathcal{O}\mathcal{A}} \dots \times_{\mathcal{O}\mathcal{A}} \mathcal{M}\mathcal{A}, \text{ q factors } \mathcal{M}\mathcal{A};$$

$$\partial_0 = S \text{ and } \partial_1 = T \text{ on } B_1\mathcal{A}; \quad s_0 = I \text{ on } B_0\mathcal{A};$$

$$\partial_i[f_1, \ldots, f_q] = \begin{cases} [f_2, \ldots, f_q] & \text{if } i = 0 \\ [f_1, \ldots, f_{i-1}, f_i f_{i+1}, \ldots, f_q] & \text{if } 0 < i < q \text{ for } q > 1; \\ [f_1, \ldots, f_{q-1}] & \text{if } i = q \end{cases}$$

$$s_i[f_1, \ldots, f_q] = [f_1, \ldots, f_i, \mathrm{ISf}_i = \mathrm{ITf}_{i+1}, f_{i+1}, \ldots, f_q] \text{ for } q > 0.$$

Define the classifying space $B\mathcal{A}$ of $\mathcal{A}$ to be the geometric realization of $B_*\mathcal{A}$. Then B is a functor from the category of (topological) categories and functors to the category of spaces and maps; B preserves products by [G, 11.5]. Let $\mathcal{J}$ denote the category with objects 0 and 1 and a single non-identity morphism $0 \to 1$; $B\mathcal{J}$ is homeomorphic to the unit interval I. A natural transformation $\lambda : F \to G$ between functors $\mathcal{A} \to \mathcal{A}'$ determines a functor $\lambda : \mathcal{A} \times \mathcal{J} \to \mathcal{A}'$ and thus a homotopy $B\lambda : B\mathcal{A} \times I \to B\mathcal{A}'$ between BF and BG.

In [G, §9 and §10], the bar construction was set up in sufficient generality to ensure that anything which looks like a bar construction is indeed a bar construction. The above construction is no exception. A category with object space $\mathcal{O}$ is a monoid in the monoidal category of (topological) $\mathcal{O}$-Graphs [14, p. 49 and 167], and $B_*\mathcal{A} = B_*(\mathcal{O}, \mathcal{A}, \mathcal{O})$ where $\mathcal{O}$ is the identity object of $\mathcal{O}$-Graph. When $\mathcal{O}$ is a single point $*$, $\mathcal{O}$-Graph is the category $\mathcal{U}$ of spaces and the present functor B reduces to the classifying space functor for topological monoids. We also have the following familiar special case, the normalized version of Milnor's universal bundle for topological groups.

Lemma 4.7. <u>For a topological group G, $B\tilde{G} = |D_*G|$ where D_*G is as defined in [G, 10.2]. Therefore $B\tilde{G}$ and EG are homeomorphic as right G-spaces.</u>

Proof. $D_qG = G^{q+1}$ with faces and degeneracies given by projections and diagonals. A simplicial homeomorphism $B_*\tilde{G} \leftrightarrow D_*G$ is given by $g \leftrightarrow g$ on 0-simplices and by

$$[g_1 \leftarrow g_2, \ldots, g_q \leftarrow g_{q+1}] \leftrightarrow (g_1, \ldots, g_{q+1})$$

on q-simplices for $q > 0$. The second statement follows from [G, 10.3].

An equally trivial comparison of definitions gives the following addendum (compare 4. 4 to the proof of 3. 7).

Lemma 4. 8. *The* E_∞ *operad* $\mathcal{D}$ *of* [G, p. 161] *satisfies* $\mathcal{D}(j) = B\tilde{\Sigma}_j$ *as a right* Σ_j*-space and its structural maps* γ *coincide with the maps* $B\tilde{\gamma} : B\tilde{\Sigma}_k \times B\tilde{\Sigma}_{j_1} \times \ldots \times B\tilde{\Sigma}_{j_k} \to B\tilde{\Sigma}_j$, $j_1 + \ldots + j_k = j$.

Now, by the very definition of an action by an operad [G, 1. 4], 4. 3 and 4. 4 immediately imply the following result.

Theorem 4. 9. *If* $(\mathcal{A}, \square, *, c)$ *is a permutative category, then the maps*

$$\Gamma_j = Bc_j : \mathcal{D}(j) \times (B\mathcal{A})^j \to B\mathcal{A}$$

define a natural action Γ *of the* E_∞ *operad* $\mathcal{D}$ *on* $B\mathcal{A}$.

Similarly, 4. 5 implies the following consistency statement.

Theorem 4. 10. *If* $(\mathcal{A}, \square, *)$ *is a strict monoidal category, then* $B\mathcal{A}$ *is a topological monoid with product* $B\square$. *If* $\mathcal{A}$ *is permutative, then* $(B\mathcal{A}, \Gamma, B\square)$ *is a homotopy monoid in* $\mathcal{D}[\mathcal{T}]$ *in the sense that the following diagrams are* Σ_j*-equivariantly homotopy commutative,* $j \geq 0$:

$$\begin{array}{ccc}
\mathcal{D}(j) \times (B\mathcal{A} \times B\mathcal{A})^j & \xrightarrow{(\Gamma \times \Gamma)_j} & B\mathcal{A} \times B\mathcal{A} \\
\downarrow{\scriptstyle 1 \times (B\square)^j} & & \downarrow{\scriptstyle B\square} \\
\mathcal{D}(j) \times (B\mathcal{A})^j & \xrightarrow{\Gamma_j} & B\mathcal{A}
\end{array}$$

In particular, $B\square$ *is homotopic to* $\theta_2(d)$ *for any* $d \in \mathcal{D}(2)$.

Proof. For the last statement, merely replace equalities by homotopies in the proof of 3. 4.

In the spirit of Quillen's work [23, 24], we suggest the following as a reasonable construction of algebraic K-theory.

Definition 4. 11. For a space X, a permutative category $\mathcal{A}$, and an integer n, define $K^n(X; \mathcal{A}) = [X, B_n B\mathcal{A}]$, where B_n is the n-th de-looping functor if $n \geq 0$ and $B_{-n} = \Omega^n B_0$ if $n > 0$. In particular, define the K-groups of $\mathcal{A}$ by $K^n\mathcal{A} = \pi_0 B_n B\mathcal{A}$.

For $n \geq 0$, $K^{-n}\mathcal{A} = \pi_n B_0 B\mathcal{A}$. We have that $K^0\mathcal{A}$ is the group completion of $\pi_0 B\mathcal{A}$ where, by [G, 11.11], $\pi_0 B\mathcal{A}$ is the quotient monoid obtained from $\pi_0 \mathcal{O}\mathcal{A}$ by identifying two components [A] and [A'] whenever there is a morphism between A and A'. Thus, if all morphisms of $\mathcal{A}$ are isomorphisms and $\mathcal{O}\mathcal{A}$ is discrete, then $\pi_0 B\mathcal{A}$ is the monoid of isomorphism classes of objects of $\mathcal{A}$ and $K^0\mathcal{A}$ is the obvious Grothendieck group.

Now let R be a (topological) ring with unit. By an abuse of terminology justified by 4.2, let $\mathcal{F} \subset \mathcal{P}$ denote permutative categories under $\oplus$ obtained by choosing skeletons of the categories of finitely generated free and finitely generated projective left R-modules and their isomorphisms. Here $\mathcal{O}\mathcal{P}$ is given the discrete topology, but Aut P for $P \in \mathcal{O}\mathcal{B}$ and therefore $\mathcal{M}\mathcal{P} = \coprod_{P \in \mathcal{O}\mathcal{P}} \operatorname{Aut} P$ are assumed to be appropriately topologized. (For instance, we could topologize free, hence also projective, modules in the evident way and then use the compact-open topology.) We assume that each Aut P has the homotopy type of a CW-complex. Clearly

$$H_*B\mathcal{F} = \coprod_{n \geq 0} H_*BGL(R, n) \quad \text{and} \quad H_*B\mathcal{P} = \coprod_{P \in \mathcal{O}\mathcal{P}} H_*B \operatorname{Aut} P.$$

For example, if R is the real numbers (resp., complex numbers) and if GL(R, n) is topologized as usual, then $K^n(X; \mathcal{F})$ is real (resp., complex) connective K-theory. (For $n > 0$, this requires an easy consistency argument based on the fact that the iterated Bott maps define morphisms of permutative categories; the details are similar to those in [18, §6] where a different, more geometric, construction of these K-theories within the context of E_∞ spaces is given.)

We use $\mathcal{P}$ to define the K-theory of R; that is, we set $K^nR = K^n\mathcal{P}$. For $n < 0$, we could just as well use $\mathcal{F}$. Indeed, the translation functor (defined in the proof of 3.9) for $B\mathcal{F}$ is clearly cofinal with that for $B\mathcal{P}$ and the map $(B_0 B\mathcal{F})_0 \to (B_0 B\mathcal{P})_0$ of base-point components is therefore a homotopy equivalence since it is a map of connected H-spaces of the homotopy type of CW-complexes (by A.3,

A. 6, and [21]) which induces an isomorphism on homology.

Recall that Bass defines $K_1R = GLR/ER$ and Milnor defines $K_2R = \mathrm{Ker}(STR \to GLR)$ where ER denotes the commutator subgroup of GLR and STR denotes the Steinberg group. Provided that GLR and its topological subgroup ER are discrete, it follows that $K_1R \cong H_1BGLR$ and $K_2R \cong H_2BER$. In our general topological situation, we have the following result.

Proposition 4.12. K^0R _is isomorphic to the projective class group of_ R, $K^{-1}R$ _is isomorphic to_ H_1BGLR, _and, if the homogeneous space_ GLR/ER _is discrete,_ $K^{-2}R$ _is isomorphic to_ H_2BER.

Proof. We have already verified the first statement and, by the following argument, essentially due to Anderson (see Quillen [24]), the other two statements are consequences of 3.9. By 3.8 and 4.10, $\overline{B\mathfrak{F}}$ has the homotopy type of $BGLR$ and therefore, by 3.9, the evident isomorphism from H_*BGLR to $H_*(B_0B\mathfrak{F})_0$ is induced by $\overline{\iota}:\overline{B\mathfrak{F}} \to (B_0B\mathfrak{F})_0$. Since $\pi_1(B_0B\mathfrak{F})_0$ is Abelian,

$$K^{-1}R \cong \pi_1(B_0B\mathfrak{F})_0 \cong H_1(B_0B\mathfrak{F})_0 \cong H_1\overline{B\mathfrak{F}} \cong H_1BGLR.$$

Let $E(R, n)$ be the commutator subgroup of $GL(R, n)$. Regard $\mathfrak{F}$ as the category with objects $N = \{n \mid n \geq 0\}$ whose only morphisms are $\mathfrak{F}(n, n) = GL(R, n)$. Let $\mathcal{E}$ and $\mathfrak{F}/\mathcal{E}$ be the categories with objects N whose only morphisms are

$$\mathcal{E}(n, n) = E(R, n) \quad \text{and} \quad (\mathfrak{F}/\mathcal{E})(n, n) = GL(R, n)/E(R, n).$$

Let $i : \mathcal{E} \to \mathfrak{F}$ and $\pi : \mathfrak{F} \to \mathfrak{F}/\mathcal{E}$ denote the evident morphisms of permutative categories under $\oplus$ (and cognate maps). We then have the following homotopy commutative diagram:

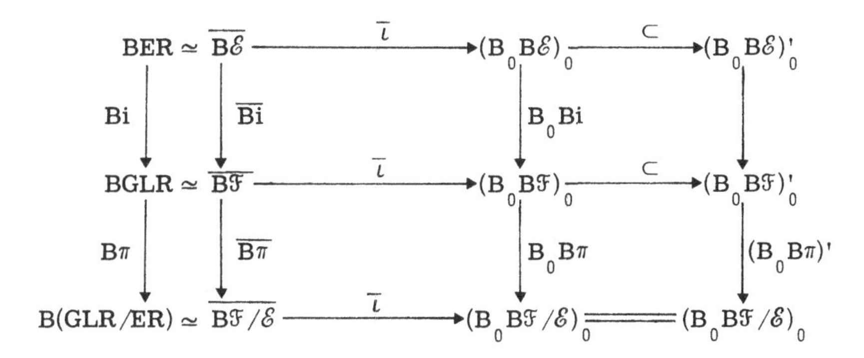

On the right, $(B_0B\pi)'$ if the fibration with fibre $(B_0B\mathcal{E})'_0$ obtained by replacing $B_0B\pi$ by a fibration in the standard fashion; $(B_0B\pi)'$ is an H-map and therefore has trivial local coefficients in homology [see 9, p. 16.08, 16.09]. As noted by Quillen [24, p. 18], $B\pi$ on the left is a fibration with trivial local coefficients because if $x = \pi(y) \in GLR/ER$ and $z \in H_*BER$, then there is a conjugate subgroup G of $E(R, n)$ in ER for some n such that z is in the image of H_*BG and y commutes with the elements of G, hence x acts trivially on z. Observe that, by suitably expanding the diagram, following the proof of 3.9, we could arrange to have all squares commute. We may thus regard the diagram as a map of fibrations. It follows by the comparison theorem that $H_*BER \to H_*(B_0B\mathcal{E})'_0$ is an isomorphism. Since GLR/ER is discrete and Abelian, $B(GLR/ER)$ is a $K(GLR/ER, 1)$ and an Abelian H-space. It is easily verified by consideration of the relevant limits that $\bar{\iota} : \overline{B\mathcal{F}/\mathcal{E}} \to (B_0\mathcal{F}/\mathcal{E})_0$ is an H-map and therefore (since it induces an isomorphism on homology) a homotopy equivalence. Thus

$$K^{-2}R \cong \pi_2(B_0B\mathcal{F})_0 \cong \pi_2(B_0B\mathcal{E})'_0 \cong H_2(B_0B\mathcal{E})'_0 \cong H_2BER.$$

APPENDIX

In two results of [G], namely [G, 3.4 and 11.13], connectivity was assumed only because I was unaware of the following fundamental 'glueing theorem' of R. Brown [8; 7.5.7].

Theorem A.1. Suppose given a commutative diagram

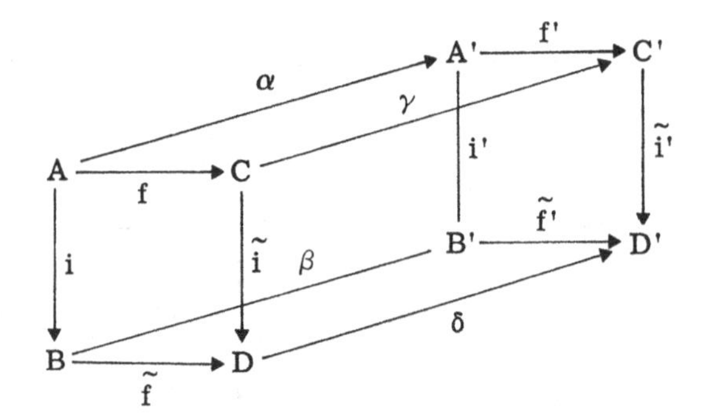

of spaces and maps such that i *and* i' *are cofibrations and* $\tilde{i}$ *and* $\tilde{i}'$ *are the cofibrations induced by* f *and* f'. *(Thus the front and back squares are pushouts.) Assume that* α, β, *and* γ *are homotopy equivalences. Then* δ *is also a homotopy equivalence.*

The second part of the following sharpening of [G, 3.4] was known to Beck; it is closely related to his result [5, Theorem 8] on topological theories.

Proposition A.2. *Let* $\psi : \mathcal{C} \to \mathcal{C}'$ *be a morphism of operads and let* $X \in \mathcal{T}$.

(i) *If* ψ *is a local equivalence and* $\mathcal{C}$ *and* $\mathcal{C}'$ *are* Σ*-free, then* $\psi : CX \to C'X$ *induces an isomorphism on (integral) homology.*

(ii) *If* ψ *is a local* Σ*-equivalence, then* $\psi : CX \to C'X$ *is a homotopy equivalence.*

Proof. The hypotheses are explained in [G, p. 21] and (i) is proven (but not stated) in [G, p. 22]. CX is constructed by means of successive pushouts

$$\begin{array}{ccc} \mathcal{C}(j+1) \times_{\Sigma_{j+1}} sX^j & \xrightarrow{\quad f \quad} & F_j CX \\ \cap \big\downarrow & & \big\downarrow \\ \mathcal{C}(j+1) \times_{\Sigma_{j+1}} X^{j+1} & \longrightarrow & F_{j+1} CX \end{array}$$

where $sX^j = \bigcup_{0 \le i \le j} s_i X^j$, $s_i(x_1, \ldots, x_j) = (x_1, \ldots, x_{i-1}, *, x_i, \ldots, x_j)$, and $f(c, s_i y) = [\sigma_i c, y]$ for $c \in \mathcal{C}(j+1)$ and $y \in X^j$ (see [G, 2.3 and 2.4] for the notations). By induction on j, the hypothesis of (ii), the non-degeneracy of the base-point of X, and the glueing theorem imply that $\psi : F_j CX \to F_j C'X$ is a homotopy equivalence for all j. The result follows.

Corollary A. 3. Let $\mathcal{C}$ be a Σ-free operad such that $\mathcal{C}(j)$, $j \geq 1$, has the Σ_j-equivariant homotopy type of a CW-complex. Then CX has the homotopy type of a CW-complex.

Proof. Let S and T denote the total singular complex and geometric realization functors between spaces and simplicial sets and let $\Phi : TS \to 1$ be the standard natural transformation. Recall that Φ is a homotopy equivalence on spaces of the homotopy type of CW-complexes [15, 16.6]. Let $\mathcal{C}'$ be the operad $TS\mathcal{C}$ ($\mathcal{C}'(j) = TS\mathcal{C}(j)$, etc.) and let $X' = TSX$. By the freeness of the Σ_j actions, $\Phi : \mathcal{C}' \to \mathcal{C}$ is a local Σ-equivalence, hence $\Phi : C'X' \to CX'$ is a homotopy equivalence. By an argument just like the previous proof, $C\Phi : CX' \to CX$ is also a homotopy equivalence. $\mathcal{C}'(j)$ is a CW free Σ_j-complex, and it follows by induction and the glueing diagrams that $C'X'$ is a CW-complex.

The second part of the following sharpening of [G, 11.13] is due to Tornehave [28, A. 3]; we give his proof for completeness. Zisman (private communication) has an alternative proof based on a homotopy analog of the Segal [25; G, 11.14] spectral sequence in homology.

Theorem A. 4. Let $f : X \to X'$ be a map of proper simplicial spaces.

(i) If each $f_q : X_q \to X'_q$ induces an isomorphism on homology, then $|f| : |X| \to |X'|$ induces an isomorphism on homology.

(ii) If each $f_q : X_q \to X'_q$ is a homotopy equivalence, then $|f| : |X| \to |X'|$ is a homotopy equivalence.

Proof. (i) follows either from the Segal spectral sequence or from a slight refinement (see A. 5) of the proof of [G, 11.13]. We prove (ii). Since X is proper, the inclusion $sX_q \to X_{q+1}$ is a cofibration, where $sX_q = \bigcup_{0 \leq j \leq q} s_j X_q$. We have successive pushouts

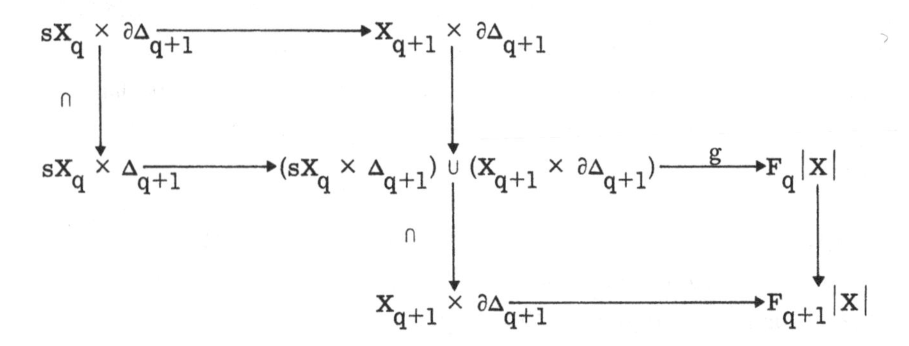

where $q(s_i x, u) = |x, \sigma_i u|$ and $q(x, \delta_i v) = |\partial_i x, v|$ (see [G, 11.1] for the notation). By induction on q and the glueing theorem, it suffices to show that $f_{q+1} : sX_q \to sX'_q$ is a homotopy equivalence for all q. Let $s^kX_q = \bigcup_{0\leq j\leq k} s_jX_q$ for $0 \leq k \leq q$. In the following diagram, the right square is a pushout and the maps s_k are homeomorphisms:

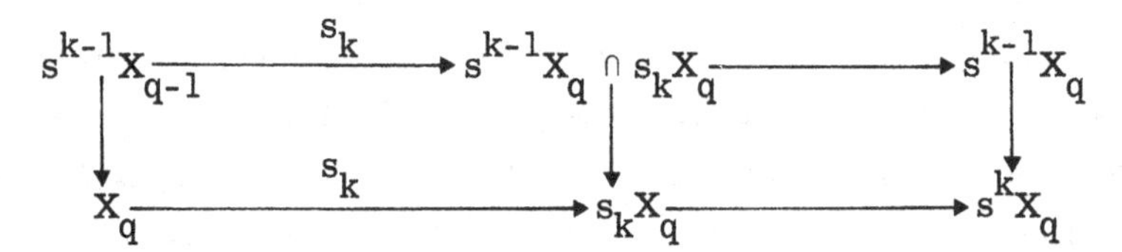

If we assume inductively that $s^{k-1}X_{q-1} \to s^kX_{q-1}$ is a cofibration for $0 < k < q$ (a vacuous assumption if $q = 0$ or 1), then we conclude (by [G, A.5] and propriety) that $s^{k-1}X_q \to s^kX_q$ is a cofibration, and similarly for X'. Since $s_0 : X_q \to s_0X_q$ is a homeomorphism, $f_{q+1} : s^0X_q \to s^0X'_q$ is a homotopy equivalence. By induction on q and for fixed q by induction on k, the diagram above and the glueing theorem imply that $f_{q+1} : s^kX_q \to s^kX'_q$ is a homotopy equivalence for all k and q. The result follows.

Remarks A.5. In the proof of [G, 11.12], which asserts that $|X|$ is n-connected if each X_q is $(n - q)$-connected, we can use the Mayer-Vietoris sequence of the excision

$$(s_kX_{q-1}, s_kX_{q-1} \cap s^{k-1}X_{q-1}) \to (s^kX_q, s^{k-1}X_q)$$

instead of that used in [G, p. 108-109] (in view of the proof above). It follows that 'strict' propriety (see [G, 11.2]) was an unnecessary hypothesis in [G, 11.12]. Since strictness was not required elsewhere, it is an unnecessary notion and references to it should be deleted throughout [G].

The following corollary is also due to Tornehave [28, A.5].

Corollary A.6. _If X is a proper simplicial space such that each X_q has the homotopy type of a CW-complex, then $|X|$ has the homotopy type of a CW-complex._

Proof. With notations as in the proof of A.3, $|\Phi_*| : |T_*S_*X| \to |X|$ is a homotopy equivalence. Since T_*S_*X is a cellular simplicial space (each ∂_i and s_i is cellular), $|T_*S_*X|$ is a CW-complex by [G, 11.4].

REFERENCES

1. D. W. Anderson. Spectra and Γ-sets. _Proceedings of Symposia in Pure Mathematics_, Vol. 22. Amer. Math. Soc. 1971, 23-30.
2. D. W. Anderson. _Simplicial K-theory and generalized homology theories I, II_ (Preprint).
3. M. G. Barratt. A free group functor for stable homotopy. _Proceedings of Symposia in Pure Mathematics_, Vol. 22. Amer. Math. Soc. 1971, 31-6.
4. M. G. Barratt and S. Priddy. On the homology of non-connected monoids and their associated groups. _Comm. Math. Helv._ 47 (1972), 1-14.
5. J. Beck. On H-spaces and infinite loop spaces. _Lecture Notes in Mathematics_, Vol. 99 (Springer-Verlag, 1969), 139-53.
6. J. M. Boardman. Homotopy structures and the language of trees. _Proceedings of Symposia in Pure Mathematics_, Vol. 22. Amer. Math. Soc. 1971, 37-58.
7. J. M. Boardman and R. M. Vogt. Homotopy-everything H-spaces. _Bull. Amer. Math. Soc._ 74 (1968), 1117-22.
8. R. Brown. _Elements of Modern Topology_ (McGraw-Hill, 1968).
9. H. Cartan. _Séminaire Henri Cartan 1959/60 exposé 16_ (École Normale Supérieure, 1961).

10. F. Cohen. Ph.D. Thesis. University of Chicago, 1972.

11. E. Dyer and R. K. Lashof. Homology of iterated loop spaces. Amer. J. Math. 84 (1962), 35-88.

12. J. R. Isbell. On coherent algebras and strict algebras. J. Algebra, 13 (1969), 299-307.

13. S. MacLane. Natural associativity and commutativity. Rice University Studies, 49 (1963), 28-46.

14. S. MacLane. Categories for the Working Mathematician (Springer-Verlag, 1971).

15. J. P. May. Simplicial Objects in Algebraic Topology (Van Nostrand, 1967).

16. J. P. May. Homology operations on infinite loop spaces. Proceedings of Symposia in Pure Mathematics, Vol. 22. Amer. Math. Soc. 1971, 171-86.

17. J. P. May. The Geometry of Iterated Loop Spaces. Lecture Notes in Mathematics, Vol. 271 (Springer-Verlag, 1972).

18. J. P. May. The Homology of E_∞ Spaces and Infinite Loop Spaces (in preparation).

19. J. P. May. Classifying spaces and fibrations (in preparation).

20. R. J. Milgram. Iterated loop spaces. Annals of Math. 84 (1966), 386-403.

21. J. Milnor. On spaces having the homotopy type of a CW-complex. Trans. Amer. Math. Soc. 90 (1959), 272-80.

22. S. B. Priddy. On $\Omega^\infty S^\infty$ and the infinite symmetric group. Proceedings of Symposia in Pure Mathematics, Vol. 22. Amer. Math. Soc. 1971, 217-20.

23. D. Quillen. Cohomology of groups. Proceedings of the International Congress of Mathematicians 1970.

24. D. Quillen. On the group completion of a simplicial monoid (preprint).

25. G. Segal. Classifying spaces and spectral sequences. Pub. Math. des Inst. des H. E. S. no. 34 (1968), 105-12.

26. G. Segal. Categories and cohomology theories (to appear).

27. J. D. Stasheff. H-spaces and classifying spaces. Proceedings of Symposia in Pure Mathematics, Vol. 22. Amer. Math. Soc. 1971, 247-72.

28. J. Tornehave. On BSG and the symmetric groups (preprint).

Department of Mathematics
The University of Chicago
5734 University Avenue
Chicago, Illinois 60637
U. S. A.

HIGHER K-THEORY FOR CATEGORIES WITH EXACT SEQUENCES

DANIEL QUILLEN

To a ring A with identity is attached a sequence of abelian groups K_iA, $i \geq 0$ which may be defined as follows. Let $\mathcal{P}_A$ be the category of finitely generated left A-modules, endowed with the direct sum operation. By work of Segal and Anderson (cf. [1]), a category with a coherent associative and commutative operation such as $\mathcal{P}_A$ determines a connected generalized cohomology theory. The groups K_iA are the coefficient groups of this cohomology theory. One can prove that they agree with the K-groups in degrees ≤ 2 introduced by Bass and Milnor (cf. [4]), and with the ones computed for a finite field in [5].

However, it is clear from the existing K-theory in low degrees that, in order to establish the basic properties of K_*A for regular rings A, one requires K-groups for the category of all finitely generated A-modules, in which the relations come from exact sequences, not just direct sums. In the present paper we outline a higher K-theory for categories with exact sequences, which enables one to prove the homotopy axiom: $K_*A = K_*(A[T])$ for regular rings, and a localization exact sequence for Dedekind domains. Full details will appear elsewhere.

§1. The space $BGL(A)^+$ and the groups K_iA. Let $f : X \to Y$ be a map of connected CW complexes with basepoint. We call f <u>acyclic</u> if the following equivalent conditions are satisfied:

(i) $H_*(X, f^*L) \xrightarrow{\sim} H_*(Y, L)$ for any local coefficient system L on Y.

(ii) The homotopy-theoretic fibre F of f is an acyclic space, i.e. $\tilde{H}_*(F, \mathbf{Z}) = 0$. (F is the space of pairs (x, p), where $x \in X$ and p is a path joining f(x) to the basepoint of Y.)

If f is acyclic, then $\pi_1(X)/N \xrightarrow{\sim} \pi_1(Y)$, where N is a normal subgroup of $\pi_1(X)$ which is perfect (equal to its commutator subgroup). Conversely, given a connected CW complex X and a perfect normal subgroup N of its fundamental group, one shows there exists an acyclic map f with source X, which is unique up to homotopy, such that N is the kernel of $\pi_1(f)$.

Now let A be a ring (supposed always to be associative with identity), let GL(A) be its infinite general linear group, and let BGL(A) be a classifying space for the discrete group GL(A). The commutator subgroup E(A) of $GL(A) = \pi_1(BGL(A))$ is perfect, so by the preceding there exists an acylic map

$$f : BGL(A) \to BGL(A)^+$$

unique up to homotopy, such that E(A) is the kernel of $\pi_1(f)$. The K-groups of the ring A are defined to be the homotopy groups of the space $BGL(A)^+$:

$$K_i A = \pi_i(BGL(A)^+) \quad \text{for} \quad i \geq 1.$$

These groups are closely connected with the homology of GL(A) and related groups, such as E(A) and the Steinberg group St(A). One has isomorphisms

$$K_1 A = H_1(GL(A), \mathbf{Z})$$
$$K_2 A = H_2(E(A), \mathbf{Z})$$
$$K_3 A = H_3(St(A), \mathbf{Z})$$

showing that the above definition agrees with the K_1 of Bass and the K_2 of Milnor. Moreover, $BGL(A)^+$ is a loop space, which has the same homology as BGL(A) as f is acyclic. Thus by a theorem of Milnor and Moore one has isomorphisms

$$K_i A \otimes \mathbf{Q} \xrightarrow{\sim} \mathcal{P}H_i(GL(A), \mathbf{Q})$$

where $\mathcal{P}$ denotes the subspace of primitive elements.

There are two basic examples where the K-groups have been calculated in all dimensions. The case of a finite field is treated in [5]. When A is the ring of S-integers in a number field, Borel [2] has determined the groups $K_* A \otimes \mathbf{Q}$. In both cases one proceeds by computing the homology of GL(A) with appropriate coefficients, using techniques special to the type of ring under consideration.

Starting from these examples, the theorems that follow may be used to produce many rings A for which the K-groups, or at least the groups $K_*A \otimes \mathbf{Q}$, can be determined.

§2. Higher K-groups for categories with exact sequences. Let $\mathcal{A}$ be a small abelian category, and let $\mathcal{M}$ denote a full subcategory of $\mathcal{A}$ closed under extensions and containing the zero object. If M is an object of $\mathcal{M}$, then by an $\mathcal{M}$-subquotient of M, we mean a quotient of the form M_2/M_1, where M_1 and M_2 are subobjects of M such that $M_1 \subset M_2$, and such that M_1, M_2/M_1, and M/M_2 are objects of $\mathcal{M}$.

We define a new category $Q(\mathcal{M})$ having the same objects as $\mathcal{M}$ in the following way. A morphism in $Q(\mathcal{M})$ from M' to M is an isomorphism of M' with an $\mathcal{M}$-subquotient of M. Such a morphism is the same as an isomorphism class of diagrams

(*)

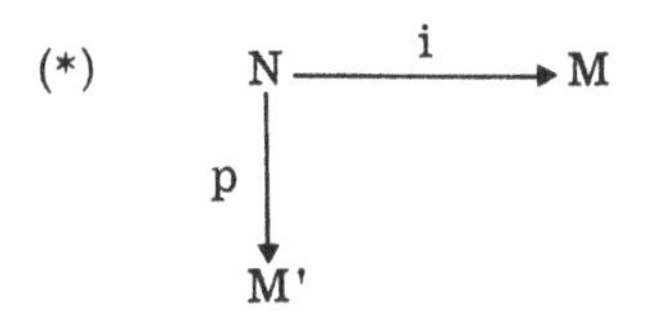

where i is a monomorphism with cokernel in $\mathcal{M}$, and p is an epimorphism with kernel in $\mathcal{M}$. The morphism in $Q(\mathcal{M})$ are composed in the evident way. Thus given a morphism from M" to M' represented by the arrows i', p' in the diagram

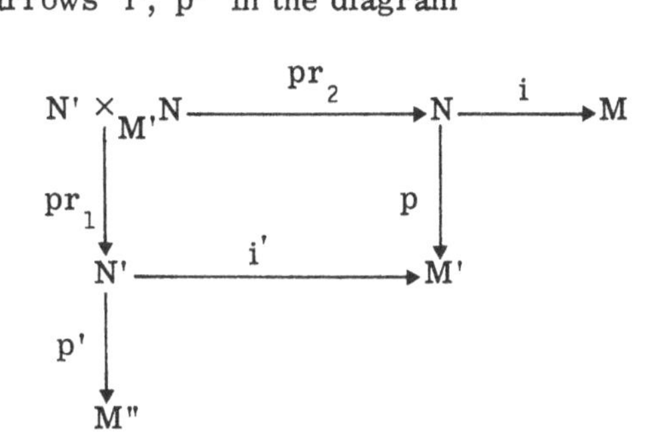

its composition with (*) is represented by the arrows $i.pr_2$ and $p'.pr_1$.

Let $|Q(\mathfrak{M})|$ denote the geometric realization of the nerve of the category $Q(\mathfrak{M})$, the nerve being the semi-simplicial set whose n-simplices are chains of composable arrows of length n. The zero object 0 may be interpreted as a basepoint of this space, hence we can make the following

Definition. $K_i(\mathfrak{M}) = \pi_{i+1}(|Q(\mathfrak{M})|, 0)$ for $i \geq 0$.

In order to make this reasonable, note that for any M in $Q(\mathfrak{M})$ there are two arrows

$$0 \rightrightarrows M$$

which result from viewing 0 as a subobject and as a quotient of M. Thus each object determines a loop in the space $|Q(\mathfrak{M})|$. Using the standard description of the fundamental group of a semi-simplicial set in terms of a maximal tree, it is not difficult to show that by means of this correspondence, the fundamental group of $|Q(\mathfrak{M})|$ is isomorphic to the Grothendieck group of the category $\mathfrak{M}$.

Observe that the category $Q(\mathfrak{M})$ depends only on $\mathfrak{M}$ and the exact sequences of objects of $\mathfrak{M}$, hence the preceding definition makes sense for any small category with a suitable notion of exact sequence. Also it is only necessary that $\mathfrak{M}$ be equivalent to a small category, in order that $|Q(\mathfrak{M})|$ be a well-defined homotopy type. For example, we can take $\mathfrak{M}$ to be the category $\mathcal{P}_A$ of finitely generated projective left modules over the ring A, with the usual notion of exact sequence for modules. In this case we have the following basic result.

Theorem 1. _The loop space of_ $|Q(\mathcal{P}_A)|$ _is homotopy equivalent to_ $K_0A \times BGL(A)^+$, _where_ K_0A _is the Grothendieck group of_ $\mathcal{P}_A$. _Consequently_ $K_i(\mathcal{P}_A) = K_iA$ for $i \geq 0$.

This theorem, and the three that immediately follow, are proved by a detailed cohomological study of categories of the form $Q(\mathfrak{M})$.

Theorem 2. _Let_ $\mathcal{P}$ _be a full subcategory of_ $\mathfrak{M}$ _such that_

(i) _For any exact sequence in_ $\mathfrak{M}$

$$(**) \qquad 0 \to M' \to M \to M'' \to 0$$

we have

(a) $M', M'' \in \mathcal{P} \Rightarrow M \in \mathcal{P}$

(b) $M \in \mathcal{P} \Rightarrow M' \in \mathcal{P}$.

(ii) *For every object* M'' *of* $\mathcal{M}$, *there exists an exact sequence* (**) *in* $\mathcal{M}$ *with* M *and* M' *in* $\mathcal{P}$.

Then the induced map $|Q(\mathcal{P})| \to |Q(\mathcal{M})|$ *is a homotopy equivalence.*

Theorem 3. *Let* $\mathcal{A}$ *be a small abelian category and let* $\mathcal{B}$ *be a full subcategory which is abelian and such that the inclusion functor is exact. Suppose also that every object of* $\mathcal{A}$ *admits a finite filtration whose quotients are objects of* $\mathcal{B}$. *Then the map* $|Q(\mathcal{B})| \to |Q(\mathcal{A})|$ *is a homotopy equivalence.*

Theorem 4. *Let* $\mathcal{A}$ *be a small abelian category, let* $\mathcal{B}$ *be a Serre subcategory, and let* $\mathcal{A}/\mathcal{B}$ *be the quotient category. Then* $|Q(\mathcal{B})|$ *is homotopy equivalent to the homotopy-theoretic fibre of the map* $|Q(\mathcal{A})| \to |Q(\mathcal{A}/\mathcal{B})|$. *Consequently, there is a long exact sequence*

$$\to K_i(\mathcal{B}) \to K_i(\mathcal{A}) \to K_i(\mathcal{A}/\mathcal{B}) \xrightarrow{\partial} K_{i-1}(\mathcal{B}) \to .$$

§3. Some applications. If A is a left noetherian ring, let $\mathrm{Modf}(A)$ denote the abelian category of finitely generated left A-modules, and set

$$G_i A = K_i(\mathrm{Modf}(A)).$$

Recall that A is called left regular if it is left noetherian and if every object of $\mathrm{Modf}(A)$ is of finite projective dimension.

Theorem 5. *If* A *is left regular, then* $K_*A \cong G_*A$.

In effect, let $\mathcal{M}_i$ be the full subcategory of $\mathrm{Modf}(A)$ consisting of modules of projective dimension $\leq i$. Theorem 2 implies that $|Q(\mathcal{M}_{i-1})|$ is homotopy equivalent to $|Q(\mathcal{M}_i)|$ for each i, hence by a limit argument it follows that $|Q(\mathcal{P}_A)| \to |Q(\mathrm{Modf}(A))|$ is a homotopy equivalence, whence the theorem.

Theorem 6. *If I is a nilpotent ideal in a left noetherian ring A, then* $G_*(A/I) \cong G_*A$.

Theorem 7. *Let $\mathcal{A}$ be a small abelian category in which every object has finite length. Then*

$$K_*(\mathcal{A}) \cong \bigoplus_{j \in J} K_* D_j$$

where $\{X_j, \ j \in J\}$ *is a set of representatives for the isomorphism classes of simple objects in* $\mathcal{A}$, *and* D_j *is the sfield* $\mathrm{End}(X_j)$.

These result by applying Theorem 3 to the inclusion $\mathrm{Modf}(A/I) \to \mathrm{Modf}(A)$, and to the inclusion of the semi-simple objects in $\mathcal{A}$.

Theorem 8. *If A is a Dedekind domain with fraction field F, then there is a long exact sequence*

$$\to K_i A \to K_i F \xrightarrow{\partial} \bigoplus_m K_{i-1}(A/m) \to K_{i-1} A \to$$

where m runs over the set of maximal ideals of A.

This follows from Theorem 4 with $\mathcal{B}$ the full subcategory of torsion modules in $\mathcal{A} = \mathrm{Modf}(A)$, together with Theorems 5 and 7.

The transfer: If A is any ring, let $\mathcal{M}_i$ be the full subcategory of the category of left A-modules consisting of those modules which admit resolutions of length $\leq i$ by objects of $\mathcal{P}_A$. Applying Theorem 2 inductively, one sees that $|Q(\mathcal{P}_A)| \to |Q(\mathcal{M}_i)|$ is a homotopy equivalence for all i. Thus if $f : A \to B$ is a ring homomorphism such that B is an object of $\mathcal{M}_i$ for some i, then restriction of scalars provides a functor $Q(\mathcal{P}_B) \to Q(\mathcal{M}_i)$, and hence gives rise to a homomorphism

$$f_* : K_i B \to K_i A.$$

§4. Graded rings, filtered rings, and the homotopy axiom.

Theorem 9. *Let* $A = \bigoplus_{n \geq 0} A_n$ *be a graded ring such that*

(i) A *is left noetherian*

(ii) A *is flat as a right* A_0*-module*

(iii) A_0 is of finite Tor-dimension as a right A-module. Let Modfgr(A) be the category of finitely generated graded left A-modules $M = \bigoplus_{n\geq 0} M_n$. Then $K_i(\mathrm{Modfgr}(A)) \cong G_i A_0 \otimes_{\mathbf{Z}} \mathbf{Z}[T]$ for all i.

Theorem 10. Let $A = \bigcup_{n\geq 0} F_n A$ be a ring with an increasing filtration such that $1 \in F_0 A$ and $F_i A . F_j A \subset F_{i+j} A$. Suppose that the associated graded ring $\mathrm{gr}(A) = \oplus F_n A/F_{n-1} A$ satisfies the hypotheses of Theorem 9. Then $G_*(F_0 A) \cong G_* A$.

Sketch of proof: Let t denote the element of degree one of the graded ring $A' = \oplus F_n A$ represented by $1 \in F_1 A$, and let $\mathcal{B}$ be the Serre subcategory of $\mathcal{A} = \mathrm{Modfgr}(A')$ consisting of modules on which t is nilpotent. Then $\mathcal{A}/\mathcal{B}$ is equivalent to Modf(A). By Theorem 3, $|Q(\mathcal{B})|$ is homotopy equivalent to $|Q(\mathcal{B}')|$, where $\mathcal{B}'$ is the subcategory of $\mathcal{B}$ consisting of $A'/tA' = \mathrm{gr}(A)$ modules, hence the exact sequence of Theorem 4 takes the form

$$\to K_i(\mathrm{Modfgr}(\mathrm{gr}\ A)) \xrightarrow{u} K_i(\mathrm{Modfgr}(A')) \to K_i(\mathrm{Modf}(A)) \to .$$

By the preceding theorem, the source and target of u are isomorphic to $G_i(F_0 A) \otimes_{\mathbf{Z}} \mathbf{Z}[T]$; one shows that u is multiplication by $T - 1$, whence the result.

As a corollary one has the first part of the following.

Theorem 11. If A is left noetherian, then

(a) $G_i A \cong G_i(A[T])$

(b) $G_i(A[T, T^{-1}]) \cong G_i A \oplus G_{i-1} A$.

When A is left regular, G_* may be replaced by K_* in this theorem. According to Gersten [3], the isomorphism $K_* A \cong K_*(A[T])$ for left regular rings signifies that the Karoubi-Villamayor K-groups coincide with the ones considered here for such rings. Here is another application of Theorem 10.

Corollary. Let $\mathfrak{g}$ be a finite dimensional Lie algebra over a field k and $U(\mathfrak{g})$ its enveloping algebra. Then $K_*(k) \cong K_*(U(\mathfrak{g}))$.

§5. Higher K-theory for schemes. If X is a noetherian scheme, let $G_*(X)$ be the K-groups of the abelian category of coherent sheaves on X, defined as in §2. Then, at least if we restrict to schemes having ample invertible sheaves, the preceding arguments permit one to define maps $f_* : G_*(X) \to G_*(Y)$ for a proper map $f : X \to Y$, (resp. $f^* : G_*(Y) \to G_*(X)$ when f is of finite Tor-dimension) with the usual properties. In addition, one has a long exact sequence

$$\to G_i(X - U) \to G_i(X) \to G_i(U) \to G_{i-1}(X - U) \to$$

when U is an open subscheme of X, the homotopy axiom:

$$G_i(X) \cong G_i(X \times_{\text{Spec } \mathbf{Z}} \text{Spec } \mathbf{Z}[T])$$

and the projective bundle theorem:

$$G_i(\mathbf{P}E) \cong G_i(X) \otimes_{K_0(X)} K_0(\mathbf{P}E)$$

where $\mathbf{P}E$ is the projective fibre bundle associated to a vector bundle E over X, and where K_0 is the Grothendieck group of vector bundles. Finally, by filtering the category of coherent sheaves on X according to the dimension of the support, one obtains a spectral sequence

$$E^1_{pq} = \bigoplus_{\dim(x) = p} K_{p+q} k(x) \Rightarrow G_{p+q}(X)$$

relating $G_*(X)$ to the K-groups of the various residue fields of the points of X, which generalizes Theorem 8.

REFERENCES

1. D. W. Anderson. K-theory, simplicial complexes, and categories. Actes, Congrès intern. math., 1970, Tome 2, 3-11.
2. A. Borel. Cohomologie réelle stable de groupes S-arithmétiques classiques (to appear).
3. S. M. Gersten. The relationship between the K-theory of Karoubi and Villamayor and the K-theory of Quillen (to appear).
4. J. Milnor. Introduction to algebraic K-theory. Annals of Math. Studies 72, Princeton Univ. Press, 1971.

5. D. Quillen. On the cohomology and K-theory of the general linear groups over a finite field (to appear).

Massachusetts Institute of Technology
Cambridge, Massachusetts 02139
U. S. A.

OPERATIONS IN STABLE HOMOTOPY THEORY

GRAEME SEGAL

If X is a space with base-point let QX denote $\Omega^\infty S^\infty X = \varinjlim_n \Omega^n S^n X$, where $\Omega^n S^n X$ is the n-fold loop-space of the n-fold reduced suspension of X. Let $\mathbf{RP}^\infty$ denote infinite-dimensional real projective space; and S^k denote the k-dimensional sphere.

The purpose of this note is to give a short proof of the following theorem of Kahn and Priddy [1]. In fact the method proves a little more, answering a question raised by Mahowald.

Theorem 1. There are maps $\alpha : Q\mathbf{RP}^\infty \to QS^0$ and $\beta : QS^0 \to Q\mathbf{RP}^\infty$ such that $\alpha\beta : QS^0 \to QS^0$ induces an isomorphism of the 2-primary components of the homotopy groups.

This theorem follows at once from the existence of certain operations in stable cohomotopy theory described in Theorem 2 below. Let $\pi_S^q(X)$ denote the stable cohomotopy of a space X, i.e. $\pi_S^q(X) = [X; \varinjlim_n \Omega^n S^{n+q}]$, where $[\ ;\ ]$ means homotopy-classes of maps (no base-points). More generally, I write $\pi_S^q(X; Y)$ for $[X; \varinjlim_n \Omega^n S^{n+q}(Y_+)]$, where Y_+ means Y with a disjoint base-point added. Notice that a map $Y_1 \times Y_2 \to Y_3$ induces $\pi_S^p(X; Y_1) \otimes \pi_S^q(X; Y_2) \to \pi_S^{p+q}(X; Y_3)$. We shall be concerned only with the case $Y = B\Sigma_n$, the classifying-space for the nth symmetric group, and with the map $B\Sigma_n \times B\Sigma_m \to B\Sigma_{n+m}$ coming from juxtaposition of permutations. For each n there is a 'forgetful' transformation of cohomology theories

$$\phi_n : \pi_S^*(\ ; B\Sigma_n) \to \pi_S^*$$

which will be defined below.

Theorem 2. *There are transformations of functors of* X

$$\theta^n : \pi_S^0(X) \to \pi_S^0(X; B\Sigma_n)$$

for $n \geq 0$, *such that*

(a) $\theta^0(x) = 1$ *for all* x,

(b) $\theta^1(x) = x$ *for all* x,

(c) $\theta^k(x+y) = \sum_{i=0}^{k} \theta^i(x)\theta^{k-i}(y)$ *for all* x, y, *where the product on the right comes from the juxtaposition map* $B\Sigma_i \times B\Sigma_{k-i} \to B\Sigma_k$,

(d) $\phi_k\theta^k(x) = x(x-1)(x-2)\ldots(x-k+1)$.

Proof that Theorem 2 implies Theorem 1. The transformations θ^2 and ϕ_2 correspond to maps $QS^0 \xrightarrow{\theta^2} Q(B\Sigma_2)_+ \xrightarrow{\phi_2} QS^0$. The composite induces the map $x \mapsto x^2 - x$ from $\pi_S^0(X)$ to itself by (d). If X is a suspension and $x \in \tilde{\pi}_S^0(X) = \ker : \pi_S^0(X) \to \pi_S^0(\text{point})$, then $x^2 = 0$, so the composite is $x \mapsto -x$, which is an isomorphism. In particular $\phi_2\theta^2$ induces an isomorphism of homotopy-groups, so is a homotopy-equivalence. But $B\Sigma_2 = \mathbf{RP}^\infty = P$, say; and $Q(P_+) = QP \times QS^0$. The map $\phi_2 : [X; QP] \times [X; QS^0] \to [X; QS^0]$ is of the form $\phi_2(x, y) = \alpha(x) + 2y$ in view of the definition of ϕ_2 below; so if $\theta^2 = \beta x \gamma : QS^0 \to QP \times QS^0$ then $\alpha\beta(x) = \phi_2\theta^2(x) + 2\gamma(x)$. Thus $\alpha\beta$ is an isomorphism on the 2-primary component if and only if $\phi_2\theta^2$ is, and Theorem 1 is proved.

The same argument proves more generally

Theorem 1'. *For any prime* p *there are maps* $\alpha : QB\Sigma_p \to QS^0$ *and* $\beta : QS^0 \to QB\Sigma_p$ *such that* $\alpha\beta$ *induces an isomorphism of the* p-*primary components of the homotopy groups.*

The only changes when p is odd are that now $\phi_p\theta^p$ is $x \mapsto x(x-1)\ldots(x-p+1)$, which is $x \mapsto (p-1)!\,x$ on homotopy groups, which is an automorphism of the p-primary component because $(p-1)! \equiv -1 \pmod p$; and that now $\phi_p(x, y) = \alpha(x) + p!\,y$, so that again the bijectivity of $\alpha\beta$ on the p-component is equivalent to that of $\phi_p\theta^p$.

Theorem 2 will be deduced from the following theorem of Barratt, Priddy, and Quillen. Let Y be a fixed space. For any space X let A(X; Y) denote the set of isomorphism-classes of pairs $(p : X' \to X; f)$, where $p : X' \to X$ is a finite covering-space, and $f \in [X'; Y]$. $A(X; Y)$ is an abelian semigroup with respect to disjoint union of the spaces X', and is a contravariant homotopy-functor of X. Barratt, Priddy, and Quillen have proved ([3], [4], [5]).

Theorem 3. <u>There is a transformation of semigroup-valued functors</u> $A(\ ; Y) \to \pi_S^0(\ ; Y)$ <u>which is universal among transformations</u> $T : A(\ ; Y) \to F$ <u>where</u> F <u>is a representable abelian-group-valued homotopy-functor and</u> T <u>is a transformation of semigroup-valued functors.</u>

We shall need to know also that the pairing $A(\ ; Y_1) \times A(\ ; Y_2) \to A(\ ; Y_1 \times Y_2)$ defined by forming the fibre-product of covering-spaces corresponds via the transformation of Theorem 3 to the pairing $\pi_S^0(\ ; Y_1) \times \pi_S^0(\ ; Y_2) \to \pi_S^0(\ ; Y_1 \times Y_2)$ mentioned earlier.

Proof of Theorem 2. Define a transformation $\theta^n : A(X) \to A(X; B\Sigma_n)$ by associating to a covering $p : X' \to X$ the pair $(p_n : X'_n \to X; f_n : X'_n \to B\Sigma_n)$, where $p_n : X'_n \to X$ is the covering whose fibre at $x \in X$ is the set of subsets of cardinal n in $p^{-1}(x)$, and f_n is the classifying-map of the principal Σ_n-bundle $P'_n \to X'_n$ whose points are the <u>ordered</u> subsets of cardinal n from the fibres of p.

Let $\theta = \prod_{n\geq 1} \theta^n : A(X) \to \prod_{n\geq 1} A(X; B\Sigma_n)$. This is clearly a homomorphism of semigroups when the composition-law on the right is $((\alpha_1, \alpha_2, \ldots), (\beta_1, \beta_2, \ldots)) \mapsto (\gamma_1, \gamma_2, \ldots)$, where $\gamma_n = \sum_{i=0}^{n} \alpha_i \beta_{n-i}$, and one defines $\alpha_0 = \beta_0 = 1$. By composition with the transformations $A(X; B\Sigma_n) \to \pi_S^0(X; B\Sigma_n)$ of Theorem 3 one gets $\theta : A(X) \to \prod_{n\geq 1} \pi_S^0(X; B\Sigma_n)$, again a homomorphism of semigroups. But now the right-hand-side is a group, and is also a representable functor of X; so by Theorem 3 θ extends to a group-homomorphism $\theta : \pi_S^0(X) \to \prod_{n\geq 1} \pi_S^0(X; B\Sigma_n)$. This proves Theorem 2 except for statement (d).

Define a homomorphism $\phi_k : A(X; B\Sigma_k) \to A(X)$ by associating to $(p : X' \to X; f : X' \to B\Sigma_n)$ the pull-back by f of the contractible covering of $B\Sigma_k$. This extends by Theorem 3 to a transformation

$\phi_k : \pi_S^0(X; B\Sigma_k) \to \pi_S^0(X)$. (Notice that the composite $\pi_S^0(X) \to \pi_S^0(X; B\Sigma_k) \xrightarrow{\phi_k} \pi_S^0(X)$, where the first map is induced by the inclusion of a point in $B\Sigma_k$, is multiplication by $k!$.) The transformations ϕ_k have the property that $\phi_{k+m}(x, y) = \binom{k+m}{m}\phi_k(x)\phi_k(y)$ for any $x \in A(X; B\Sigma_k)$, $y \in A(X; B\Sigma_m)$, because Σ_{k+m} breaks up into $\binom{k+m}{m}$ cosets of $\Sigma_k \times \Sigma_m$. By universality the same formula is true on $\pi_S^0(X; B\Sigma_k) \times \pi_S^0(X; B\Sigma_m)$. Thus $\prod_{n\geq 1} \phi_n$ is a group-homomorphism $\prod_{n\geq 1} \pi_S^0(X; B\Sigma_n) \to \prod_{n\geq 1} \pi_S^0(X)$ when the product on the right is $((\alpha_1, \alpha_2, \ldots), (\beta_1, \beta_2, \ldots)) \mapsto (\gamma_1, \gamma_2, \ldots)$, with $\gamma_n = \sum_{i=0}^{n} \binom{n}{i}\alpha_i\beta_{n-i}$. Let us compare the homomorphism $\prod_{k\geq 1} \phi_k\theta^k : \pi_S^0(X) \to \prod_{k\geq 1} \pi_S^0(X)$ with the homomorphism $x \mapsto (\eta_1(x), \eta_2(x), \ldots)$, where $\eta_k(x) = x(x-1)\ldots(x-k+1)$. The composition $\xi_k = \phi_k\theta^k : A(X) \to A(X)$ assigns to a covering $X' \to X$ the covering $P_n' \to X$ defined above. By set-theory one finds that $x^k = \sum_{i=1}^{k} c_{ik}\xi_k(x)$ in $A(X)$, where c_{ik} is the number of ways of partitioning a set with k elements into i non-empty subsets. Solving these equations in $\pi_S^0(X)$ gives $\xi_k(x) = x(x-1)\ldots(x-k+1) = \eta_k(x)$, for any $x \in A(X)$. Thus $\Pi\phi_k\theta^k$ and $\Pi\eta_k$ induce the same transformation from $A(X)$ to $\prod_{k\geq 1} \pi_S^0(X)$, and so, by the uniqueness in the universal property of Theorem 3, they are identical. This completes the proof of Theorem 2.

An unstable version of the operations

Let P_{nq} be the space of subsets of cardinal n in $\mathbf{R}^q$. Thus, for example, P_{2q} has the homotopy-type of $\mathbf{RP}^{q-1}$. One can think of the P_{nq} for fixed n as embedded in each other: $\ldots \subset P_{nq} \subset P_{n,q+1} \subset \ldots$ with $\bigcup_q P_{nq} = B\Sigma_n$. Then Theorem 2 can be generalized to

Theorem 4. _For each_ n _and_ q _there are transformations of functors_ $\theta_q^n : [\ ;\Omega^q S^q] \to \pi_S^0(\ ; P_{nq})$ _such that_

(a) _if_ $q \leq r$, _the diagram_

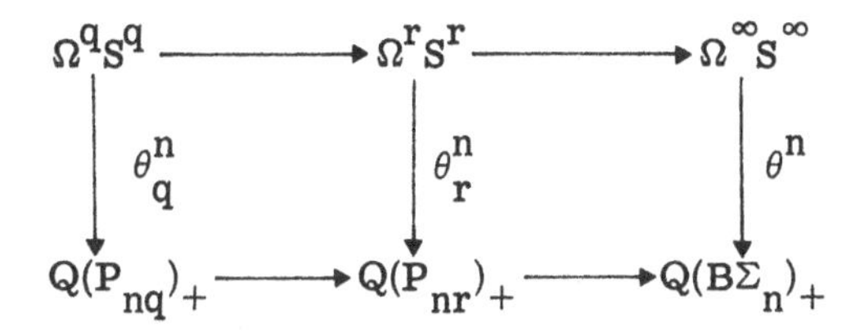

commutes, and

(b) $\theta^n_q(x + y) = \sum_{i=0} \theta^i_q(x)\,\theta^{n-i}_q(y)$, *where the pairing* $P_{iq} \times P_{n-i,q} \to P_{nq}$ *comes from any fixed open embedding* $\mathbf{R}^q \amalg \mathbf{R}^q \to \mathbf{R}^q$.

This theorem is proved by exactly the same argument as Theorem 2, but using the following generalization ([2], [6]) of the theorem of Barratt and Quillen.

Theorem 5. *Let* $A_q(X; Y)$ *be the set of homotopy-classes of objects* $(X \times \mathbf{R}^q \overset{i}{\leftarrow} X' \overset{f}{\rightarrow} Y)$, *where* i *is an embedding such that the composite* $X' \to X \times \mathbf{R}^q \to X$ *is a finite covering-map. (Two such objects are homotopic if there is an object of the same kind over the cylinder* $X \times [0, 1]$ *which restricts to them at the ends.) Then there is a transformation of semigroup-valued functors of* X $A_q(X; Y) \to [X; \Omega^q S^q(Y_+)]$ *which is universal among such transformations from* $A_q(\ ; Y)$ *into abelian-group-valued representable functors.*

REFERENCES

1. D. S. Kahn and S. B. Priddy. *Applications of the transfer to stable homotopy-theory* (to appear).
2. J. P. May. *The geometry of infinite-loop-spaces* (preprint, Cambridge).
3. S. B. Priddy. On $\Omega^\infty S^\infty$ and the infinite symmetric group. *Lecture notes of the Amer. Math. Soc. Summer Institute on Algebraic Topology, Univ. of Wisconsin* (1970).
4. D. G. Quillen. *Talk in Proceedings of Int. Cong. of Mathematicians* (Nice, 1970).
5. G. B. Segal. Categories and cohomology theories (to appear in *Topology*).

6. G. B. Segal. Configuration spaces and iterated loop-spaces Inventiones mathematicae, 20 (1973).

St Catherine's College
Oxford

EQUIVARIANT ALGEBRAIC K-THEORY

C. T. C. WALL

In a talk given three years ago in Hull, I described the foundations of equivariant algebraic K-theory (see Springer Lecture Notes, vol. 108). I will now indicate how to erect some structure on these foundations. Like the earlier account, this comes from joint work with Ali Fröhlich, and is my attempt to formalise and understand his proofs of our theorems.

The talk as given was divided into two main sections: the work on induction theorems will not be discussed here. We are concerned with the following problem: let $\mathcal{C}$ be a monoidal G-graded category, Ker $\mathcal{C}$ the subcategory of morphisms of grade 1, Rep $\mathcal{C}$ the category of representations of G by automorphisms (of the right grades) of objects of $\mathcal{C}$. What relation is there between the equivariant K-groups $K_i(\text{Rep }\mathcal{C}) = K_i(\mathcal{C}, G)$ and the ordinary ones $K_i(\text{Ker }\mathcal{C}) = K_i(\mathcal{C})$? We show that under rather special conditions, there is a spectral sequence

$$H^p(G; K_q(\mathcal{C})) \Rightarrow K_n(\mathcal{C}, G).$$

Research on this problem is still at an early stage, and the arguments are only presented in bare outline.

§1. Skeletons

The following concepts are introduced to facilitate the construction of certain simplicial sets, and verification of certain of their properties. Recall that the basic (semi) simplicial category $\mathcal{SS}$ has objects $\underline{\underline{n}} = \{0, 1, \ldots, n\}$ and morphisms increasing maps; a simplicial set is a contravariant functor $\mathcal{SS} \to \mathcal{E}ns$.

Write $\mathcal{SS}^{(n)}$ for the full subcategory of $\mathcal{SS}$ whose objects are the $\underline{\underline{r}}$ with $r \leq n$. A contravariant functor $K^{(n)} : \mathcal{SS}^{(n)} \to \mathcal{E}ns$ will be called an n-skeleton.

Lemma 1. <u>The category of simplicial sets with n-skeleton</u> $K^{(n)}$ <u>has a terminal object</u> K. <u>We can identify</u> (<u>for each</u> N) <u>the</u> N-<u>simplices of</u> K <u>with the maps into</u> $K^{(n)}$ <u>of the</u> n-<u>skeleton</u> $\Delta_N^{(n)}$ <u>of the standard</u> N-<u>simplex</u> Δ_N.

We call K the universal extension of $K^{(n)}$. Any simplicial set is said to be n-universal if it is the universal extension of its n-skeleton, and $(n - \frac{1}{2})$-universal if, in addition, any two n-simplices with the same set of faces coincide.

An n-skeleton is said to satisfy the Kan condition if (i) any set of $(r + 1)$ r-simplices $(r \leq n - 1)$, which fit together like the set of all but one of the faces of an $(r + 1)$-simplex, is indeed such a set of faces, and (ii) for any set of $(n + 1)$ n-simplices with this property, we can find a further n-simplex which fits them all.

Lemma 2. <u>If an n-skeleton is Kan, so is its universal extension.</u>

We must show, for any r, that given $(r + 1)$ r-simplices which fit together, there is an $(r + 1)$-simplex which has them as faces. For $r \leq n - 1$, this holds by (i) above. For $r = n$, we can find (by (ii) above) a further n-simplex; now by the construction of the universal extension, this set of n-simplices constitutes an $(n + 1)$-simplex. For $r > n$, any set of $(r + 1)$ r-faces of Δ_r contains the n-skeleton: the result again follows.

Corollary. <u>In this case,</u> $\pi_i(K) = 0$ <u>for</u> $i \geq n$.

Given two n-skeletons $L^{(n)}$, $K^{(n)}$, we define $\mathrm{Hom}(L^{(n)}, K^{(n)})$ to be the simplicial set, an N-simplex of which is a morphism $\Delta_N^{(n)} \times L^{(n)} \to K^{(n)}$, with the obvious face and degeneracy operators.

Lemma 3. (i) <u>Let</u> L <u>be any simplicial set</u>, K n-<u>universal.</u> <u>Then</u> $\mathrm{Hom}(L, K) = \mathrm{Hom}(L^{(n)}, K^{(n)})$.

(ii) <u>If also</u> K <u>is</u> $(n - \frac{1}{2})$-<u>universal, this</u> $\subset \mathrm{Hom}(L^{(n-1)}, K^{(n-1)})$.

(iii) $\mathrm{Hom}(L, K)$ <u>is also</u> n-<u>universal.</u>

Example 1. Let $\mathcal{C}$ be any category. Define a 2-skeleton as follows:

$S_0 = \text{ob}\,\mathcal{C}$, the objects of $\mathcal{C}$.

$S_1 = \mathcal{C}$, the morphism set of $\mathcal{C}$.

S_2 = composable pairs of morphisms of $\mathcal{C}$

(so we fill in the triangles of the form 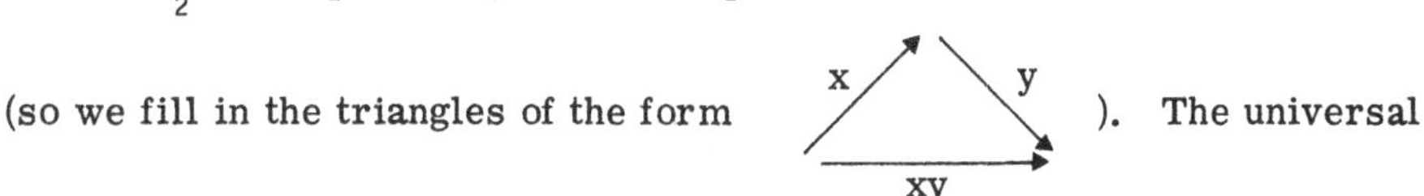

). The universal extension is the nerve $B\mathcal{C}$ of $\mathcal{C}$; clearly it is $1\frac{1}{2}$-universal. The Kan condition is equivalent to having all morphisms in $\mathcal{C}$ invertible; if this holds, $\pi_i(B\mathcal{C}) = 0$ for $i \geq 2$. It holds in particular if $\mathcal{C}$ is a group G (has just one object): we recover the usual classifying space BG.

We can construct analogously the universal covering space EG: indeed, this is the 0-universal complex with $S_0 = G$, and hence $S_i = G^{i+1}$ (the components can be regarded as the vertices of the simplex). The obvious (diagonal) action of G is free, and the quotient is identified in a natural way with BG (on the S_1 level, this is $(g_0, g_0g_1) \to g_1$).

Observe that $B(\mathcal{C} \times \mathcal{C}') = B\mathcal{C} \times B\mathcal{C}'$, and that a functor $F : \mathcal{C} \to \mathfrak{D}$ induces a morphism $BF : B\mathcal{C} \to B\mathfrak{D}$.

Example 2. (This arose in joint work with A. Fröhlich, see 'Generalisations of the Brauer group I', preprint, Liverpool University, 1971 and provided a major source of motivation for this paper.) Let R be a commutative ring. We define a $2\frac{1}{2}$-universal 3-skeleton by:

S_0 = the central separable R-algebras A_i

S_1 = invertible bimodules ${}_{A_i}M_{ij}{}_{A_j}$

S_2 = bimodule isomorphisms $f_{ijk} : M_{ij} \otimes_{A_j} M_{jk} \to M_{ik}$

S_3 = commutative diagrams

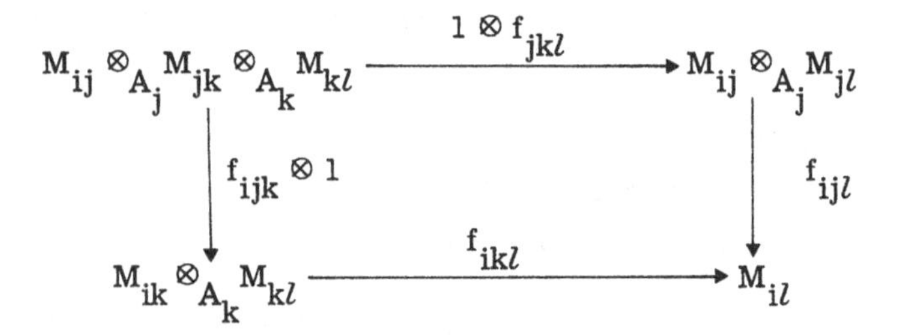

It is easy to see that the Kan condition holds here also.

§2. Algebraic K-theory

Let $\mathcal{C}$ be a monoidal category, or category-with-product in the sense of Bass. Thus we have a functor $\oplus : \mathcal{C} \times \mathcal{C} \rightarrow \mathcal{C}$, which thus induces $B\oplus : B\mathcal{C} \times B\mathcal{C} \rightarrow B\mathcal{C}$: $B\mathcal{C}$ is an H-space. In fact, $B\mathcal{C}$ is a 'homotopy everything H-space'. This can be shown in the style of Peter May as follows.

We construct a natural simplicial E_∞-operad $\mathfrak{a}$. Set $\mathfrak{a}(j) = E\sigma_j$ (where σ_j is the symmetric group on j letters). Then this is contractible, with free action of σ_j. By 0-universality, it suffices to define

$$\gamma : \ \mathfrak{a}(k) \times (\mathfrak{a}(j_1) \times \dots \times (\mathfrak{a}(j_k)) \rightarrow \mathfrak{a}(j_1 + \dots + j_k)$$

on the 0-skeleton: we take the obvious map

$$\sigma_k \times (\sigma_{j_1} \times \dots \times \sigma_{j_k}) \rightarrow \sigma_{j_1 + \dots + j_k}$$

defined by permuting the subsets of cardinals $j_1 \dots j_k$ as indicated with each other and internally. To verify the axioms, it suffices again to look at the 0-skeleton, where the verification is trivial.

To define an action of $\mathfrak{a}$ on $B\mathcal{C}$ we use the natural equivalences

$$\mathrm{com}_{A,B} : A \oplus B \rightarrow B \oplus A, \quad \mathrm{ass} : (A \oplus B) \oplus C \rightarrow A \oplus (B \oplus C)$$

and the fact that they are coherent. Thus define

$$\mathfrak{a}(j) \times (B\mathcal{C})^j \rightarrow B\mathcal{C}$$

as follows. By $(1\frac{1}{2})$-universality, it is enough to give the definition on the 1-skeleton. For the 0-skeleton,

$$\sigma_j \times (\text{ob } \mathcal{C})^j \to \text{ob } \mathcal{C} ,$$

permute the j objects in the manner indicated and take the left normalised sum

$$(\ldots((M_{\alpha(1)} \oplus M_{\alpha(2)}) \oplus M_{\alpha(3)}) \ldots \oplus M_{\alpha(j)}).$$

For the 1-skeleton

$$(\sigma_j \times \sigma_j) \times \mathcal{C}^j \to \mathcal{C}$$

suppose given permutations α, β and morphisms $f_i : M_i \to N_i$ $(1 \le i \le j)$, we add the f_i and use ass and com to obtain a morphism

$$\ldots (M_{\alpha(1)} \oplus M_{\alpha(2)}) \ldots \to \ldots (N_{\beta(1)} \oplus N_{\beta(2)}) \ldots ;$$

by coherence, the resulting morphism is uniquely determined.

Similarly, all the verifications necessary to establish that this defines an action of $\mathfrak{a}$ can be done on the 1-skeleton, and reduce to a straightforward use of coherence. Results of May now show:

Proposition. There is an Ω-spectrum $\mathcal{C}(i)$ $(i \in \mathbf{Z})$, with $\mathcal{C}(i) \simeq \Omega\mathcal{C}(i + 1)$ and $\mathcal{C}(1) = B(B(\mathcal{C}))$. If $\pi_0 B(\mathcal{C})$ is a group, $\mathcal{C}(0) \simeq B(\mathcal{C})$.

Indeed, May gives a direct construction of $\mathcal{C}(i)$ as a coend. The K-groups of the category $\mathcal{C}$ (in the sense of Quillen) are now defined as

$$K_i(\mathcal{C}) = \pi_i \mathcal{C}(0) = \pi_0 \mathcal{C}(i).$$

Note the degenerate special case when (i) all morphisms in $\mathcal{C}$ are invertible (so $B\mathcal{C}$ is Kan) and (ii) $\pi_0(B\mathcal{C})$ is a group. Here, $\mathcal{C}(0) \simeq B\mathcal{C}$ has vanishing higher homotopy groups, so $K_i(\mathcal{C}) = 0$ for $i \ge 2$. An example of interest where this occurs is when $\mathcal{C}$ is the (Picard) category of invertible R-modules and isomorphisms, for some commutative ring R, with the tensor product.

It is natural to ask in this case whether the monoidal category can be replaced by a group category, where com and ass are identity transformations and each object has an inverse. Fröhlich and I have shown that a necessary and sufficient condition for this is that for each $A \in \text{ob}\,\mathcal{C}$, $\text{com}_{A,A} : A \oplus A \to A \oplus A$ is the identity. We also show that this obstruction can be killed by redefining the equivalence com.

§3. The equivariant case

Recall that a map $p : E \to B$ of simplicial sets is a Kan fibration if for each commutative diagram

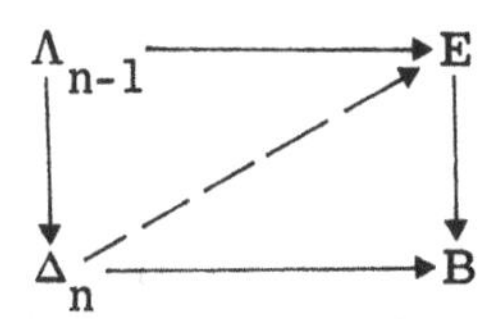

there is a map $\Delta_n \to E$ making both triangles commute. (Here, Λ_{n-1} denotes the union of all but one face of Δ_n.) If E is Kan, we can make the upper triangle commute; if B is $(n - 1\frac{1}{2})$-universal, the lower one will necessarily also commute. Now a functor $F : \mathcal{C} \to \mathcal{D}$ induces $BF : B\mathcal{C} \to B\mathcal{D}$; it follows that this is a fibration if

(i) $B\mathcal{C}$ is Kan, i.e. all morphisms in $\mathcal{C}$ are invertible

(ii) $(n = 1)$ F is a fibration, i.e. for any object C of $\mathcal{C}$ and morphism β of $\mathcal{D}$ with domain FC there is a morphism α of $\mathcal{C}$, with domain C and $F\alpha = \beta$.

(iii) $(n = 2)$ $B\mathcal{D}$ is Kan.

For G a group, a G-graded category is a pair $(\mathcal{C}, \gamma)$ where $\gamma : \mathcal{C} \to G$ is a functor. In the 'Foundations' paper, we called $\mathcal{C}$ stable if γ is a fibration. If also $B\mathcal{C}$ is Kan, we see that $B\gamma$ is then a fibration. The fibre is also $1\frac{1}{2}$-universal, and we can identify it with B(Ker $\mathcal{C}$), where Ker $\mathcal{C}$ is the subcategory of morphisms of grade 1. We are also interested in the (simplicial) set Γ of sections of this fibration. Using Lemma 3, we see again that this is 2-universal, and that it is enough to check on the 2-skeleton. Thus $\Gamma = B(\text{Rep}(\mathcal{C}, \gamma))$, where $\text{Rep}(\mathcal{C}, \gamma)$ is the 'representation category', $\text{Hom}_G(G, \mathcal{C})$.

A monoidal G-graded category has in addition a product

$$\oplus : \mathcal{C} \times_G \mathcal{C} \to \mathcal{C}$$

and again natural equivalences com and ass. Arguing just as in the preceding section, we see that we have a fibrewise action of $\mathfrak{a}$ on $B\mathcal{C}$ - i.e. if $B\mathcal{C}^{[n]}$ denotes the n-fold product of $B\mathcal{C}$ with itself over BG, then we have maps

$$\mathfrak{a}(n) \times B\mathcal{C}^{[n]} \to B\mathcal{C} ,$$

also over BG, defining actions of $\mathfrak{a}$ on each fibre.

Since May's construction is natural, we conclude that there is a sequence of fibrations

$$\mathcal{K}(i) \to \mathcal{E}(i) \to BG,$$

with $\mathcal{K}(i) \simeq \Omega\mathcal{K}(i + 1)$, and indeed each fibration has a preferred section and leads to the next by taking the fibrewise loop space. Also $\mathcal{K}(1) \simeq B(B(\mathrm{Ker}\ \mathcal{C}))$, and if $\pi_0 B(\mathrm{Ker}\ \mathcal{C})$ is a group, $\mathcal{K}(0) \simeq B(\mathrm{Ker}\ \mathcal{C})$.

In this situation there is a spectral sequence, due to Alan Robinson (thesis, University of Warwick, 1970). Write $\Gamma(i)$ for the space of cross-sections of $\mathcal{E}(i)$. There are homotopy equivalences

$$\Gamma(i) \simeq \Omega\Gamma(i + 1).$$

Then the spectral sequence has

$$E_2^{pq} = H^p(BG; \pi_q\mathcal{K}(0)) = H^p(G; K_q(\mathrm{Ker}\ \mathcal{C})),$$

labelled so that $d_r^{pq} : E_r^{p,q} \to E_r^{p+r,q+r-1}$, and the abutment is $\pi_n(\Gamma) = \pi_{n+i}\Gamma(i)$ for $n + i \geq 0$. In fact, this is filtered, and

$$F^p\pi_n(\Gamma)/F^{p+1}\pi_n(\Gamma) = E_\infty^{p,n+p} .$$

In the special case when $\pi_0 B(\mathrm{Ker}\ \mathcal{C})$ is a group, we can identify (up to homotopy) the fibrations

$$\mathcal{K}(0) \to \mathcal{E}(0) \to BG \quad \text{and} \quad B(\mathrm{Ker}\ \mathcal{C}) \to B\mathcal{C} \to BG$$

and hence the spaces of sections $\Gamma(0) \simeq B(\mathrm{Rep}\ \mathcal{C})$.

Theorem. If $\mathcal{C}$ is a monoidal G-graded category, with all morphisms invertible, and $\pi_0 B(\mathrm{Ker}\ \mathcal{C})$ is a group, there is a spectral sequence with $E_2^{p,q} = H^p(G; K_q(\mathrm{Ker}\ \mathcal{C}))$ and abutment $K_n(\mathrm{Rep}\ \mathcal{C})$ (filtered, associated to the graded group $E_\infty^{p,n+p}$).

The situation is less clear in general. The natural map $B\mathcal{C} \to \mathcal{E}(0)$ of fibrations over BG induces a map of sections, $B(\mathrm{Rep}\ \mathcal{C}) \to \Gamma(0)$, which is also a morphism of E_∞-spaces. Thus if $\mathfrak{D}(i)$ is the Ω-spectrum associated to $B(\mathrm{Rep}\ \mathcal{C})$, we have a morphism $\mathfrak{D} \to \Gamma$ of spectra. The trouble is that we lack information on this morphism. All we have shown is that there is a spectral sequence as above and a map of $K_*(\mathrm{Rep}\ \mathcal{C})$ to its abutment. To extend the full result, we need $\mathfrak{D} \to \Gamma$ to be a homotopy equivalence. It would suffice to solve affirmatively the

Problem. Suppose given a fibration with fibre a topological monoid. Do the following operations commute (up to homotopy)? -

(i) Form the space (monoid) of all sections.

(ii) Form the (fibrewise) classifying space.

It seems more probable, however, that the answer is negative.

The spectral sequence of the theorem degenerates to a Gysin sequence since, by an earlier remark, $K_q(\mathrm{Ker}\ \mathcal{C})$ vanishes unless $q = 0$ or 1. However, its construction is more general (for any 'E_∞-fibration'). The complex of §1, Example 2 can also be extended to the equivariant case, and all our conditions verified for it. Here, the homotopy groups of the fibre vanish except for $q = 0, 1, 2$, so the spectral sequence can be presented as a 'braid' diagram of 4 interlocking exact sequences. The details of this application will be presented elsewhere.

Department of Pure Mathematics
The University of Liverpool
P. O. Box 147, Liverpool, L69 3BX

ON RANK 2 MOD ODD H-SPACES

A. ZABRODSKY

0. **Introduction.** In [1], [2] and [7] the classification of rank 2 H-spaces was practically completed. The fact that there are only finitely many (homotopy classes of) such spaces is due primarily to the fact that the number of possible mod-2 cohomoloty rings of such spaces is very limited as the Steenrod algebra (as well as higher order operations) act on such rings in a very definite way.

In [4] it was shown that modulo odd primes non classical mod-p H-spaces of rank 2 exist: i.e. there exists an H-space X such that $H^*(X, Z_p)$ is an exterior algebra on two generators. (Modulo odd primes we consider odd dimensional spheres as classical H-spaces.)

Consequently, the mod odd classification of mod odd rank 2 H-spaces is yet to be considered. Its implication for the classification problem of finite CW complexes is obvious, as the corollary to Theorem A illustrates.

Throughout this paper let n and m be fixed odd integers such that $3 \leq n \leq m$. As $S^n \times S^m$ is a mod odd H-space there are mod odd H-spaces of type (n, m) for any given such pair, and the classification has to be carried out separately for each such pair.

The main contribution of this note is the following:

Theorem A. _Let_ n, m _be odd integers_ $3 \leq n \leq m$.

(a) _Let_ X _be a mod odd H-space of rank_ 2 _and of type_ n, m, i.e. _for every odd prime_ p, X _is mod_ p _equivalent to an H-space, and_ $H^*(X, Z_p) = \Lambda(x, y)$, $\dim x = n$, $\dim y = n$. _Then_ X _is mod odd equivalent to an_ S^n _fibration over_ S^m (i.e. _there exist_ $\phi : X \to E$, $r : E \to S^m$, _where_ $H^*(\phi, Z_p)$ _is an isomorphism for every odd_ p _and_ r _is a fibration with fiber_ S^n).

(b) _Let_ $\alpha \in \pi_{m-1}(S^n)$ _be odd order. Among all mod odd types of the total spaces of the_ S^n _fibrations over_ S^m _that are induced by_

α (i. e. those having $S^n \cup_\alpha e^m$ as the m-skeleton) at most one is a mod odd H-space.

(c) If $m \leq 3n$ then for every $\alpha \in \pi_{m-1}(S^n)$ of odd order there exists a mod $p > 3$ H-space among the total spaces of the S^n fibrations over S^m induced by α.

Corollary. Let X be a finite CW H-space and let p be a prime, $p > 3$. Suppose

$$H^*(X, Z_p) = \Lambda(x_{n_1}, x_{n_2}, \ldots, x_{n_k}) \quad n_1 \leq n_2 \leq \ldots \leq n_k.$$

For any pair s, t such that $n_s \leq n_t \leq 3n_s$ and for any spherical operation $\Phi \in H^{n_t}(S^{n_s}(n_t - 2), Z_p)$ (where $S^{n_s}(n_t - 2)$ is the Postnikov approximation of S^{n_s} in $\dim \leq n_t - 2$) there exists an H-space X_Φ such that $H^*(X, Z_p) \approx H^*(X_\Phi, Z_p)$ as algebras and $x_{n_t} \in \Phi(x_{n_s})$.

It follows that there are very limiting restrictions on the way Z_p $(p > 3)$ operations may act on the cohomology of a finite CW H-space. Theorem A(c) can be generalized (using an identical proof) to show that any S^n fibration over S^m is a mod p H-space provided $(n + 1)p - 3 > 2n + m$. (This is a weaker limitation on m in comparison with the formula in [4]: $2(n + m) < p(n + 1) - 2$. In particular $B_n(5)$ is a mod 5 H-space for $n \geq 3$.)

The proof of Theorem A(a) and (b) is based on the following theorem:

Theorem B. Let X, Y, Y_0 be H-spaces and p be a prime. Suppose $H^*(X, Z_p) = \Lambda(x_{n_1}, x_{n_2}, \ldots, x_{n_k})$, $3 \leq n_1 \leq n_2 \leq \ldots \leq n_k$, and $k < p$. Let $g : Y \to Y_0$ be an H-map whose fiber is n_k-connected and has finite homotopy groups of order a power of p.

Suppose $f : X \to Y_0$ satisfies

$$(b_0) \quad f_0 \circ P_{\lambda, X} \sim P_{\lambda, Y_0} \circ f_0 ,$$

where $P_{\lambda, X}$ and P_{λ, Y_0} are the λ-th power maps in X and Y_0 respectively, and λ is an integer representing a primitive root of unity in

Z_p (i.e. $\lambda^k \not\equiv \lambda(p)$ for $1 < k < p$). Then f_0 can be lifted to $f : X \to Y$, $g_0 f = f_0$ and

(b) $f \circ P_{\lambda, X} \sim P_{\lambda, Y} \circ f$.

(Note that if f_0 is an H-map (b_0) holds.) In view of Corollaries 1.4 and 1.5 one hopes to prove Theorem B without the restriction $k < p$.

Theorem B is proved in section 1. Theorem A parts (a) and (c) are proved in section 2. In section 3 we review and carry further the study of some homotopy associativity obstructions needed for the proof of Theorem A(c) in section 4.

In the proof we occasionally use some of the notations and techniques of mixing and of P_1 factorizations of rational equivalences as in [6] and [7].

1. Proof of Theorem B. One may obviously assume that $p > 2$. Let (A, ϕ, ψ) be a commutative associative Hopf algebra over Z_p. Define $P_\lambda : A \to A$ by

$$P_\lambda = \phi(\phi \otimes 1) \dots (\phi \otimes \underbrace{1 \otimes \dots \otimes 1}_{\lambda - 2})(\psi \otimes \underbrace{1 \dots \otimes 1}_{\lambda - 2}) \dots (\psi \otimes 1)\psi .$$

If $A = H^*(X, Z_p)$, where X is an H-space, then $P^*_{\lambda, X} = P_\lambda$, where $P_{\lambda, X}$ is the λ-th power map in X.

1.1. Lemma. Let (A, ϕ, ψ) be a Hopf algebra. Suppose $A = \Lambda(x_{n_1}, x_{n_2}, \dots x_{n_k})$ as an algebra ($n_i = \deg x_{n_i}$ is odd), and $n_1 \le n_2 \dots \le n_k$, $k < p$. Then $(P_\lambda - \lambda . 1) : A \to A$ induces an isomorphism of the submodule of decomposables in A onto itself.

Proof. Suppose first that x_{n_i} are primitive. Then if w is a monomial of length t, $1 < t < p$, then $P_\lambda w = \lambda^t w$ and $(P_\lambda - \lambda . 1) w = (\lambda^t - \lambda) w$. As $\lambda^t - \lambda \not\equiv 0$ (p), $P_\lambda - \lambda . 1$ is an isomorphism on the decomposables. If A is not primitively generated $(P_\lambda - \lambda . 1) w \equiv (\lambda^t - \lambda) w$ modulo monomials of length greater than t, and 1.1 follows.

1.2. Proof of Theorem B. It suffices to prove theorem B for the case where $g: Y \to Y_0$ is a $K(Z_p, m)$ principal fibration, with $m > n_p$, induced by an H-map $h: Y_0 \to K(Z_p, m+1)$. (Any other g can be factored into such maps, and the lifting process can be carried out inductively.)

Let $l_{m+1} \in H^{m+1}(K(Z_p, m+1), Z_p)$ be the fundamental class. Then $h^* l_{m+1} = v$ is primitive. Hence, by (b_0), $0 = P^*_{\lambda, X} f_0^* v - f_0^* P_{\lambda, Y_0} v = (P^*_{\lambda, X} - \lambda.1) f_0^* v$. As $\dim f_0^* v > n_k$, $f_0^* v$ is decomposable and hence, by 1.1, $(P^*_{\lambda, X} f_0^* v - \lambda f_0^* v) = 0$ implies $f_0^* v = 0$, so that f_0 can be lifted to $f': X \to Y$. One need only prove (b).

Consider $w = [f' \circ P_{\lambda, X}] - [P_{\lambda, Y} \circ f'] \in [X, Y]$. As $g_* w = [f_0 \circ P_{\lambda, X}] - [P_{\lambda, Y_0} \circ f_0] = 0$, $w = j_* \tilde{w}$ for some $\tilde{w} \in [X, K(Z_p, m)] = H^m(X, Z_p)$ where $j: K(Z_p, m) \to Y$ is the inclusion of the fiber of g. As $m > n_k$, $\tilde{w}$ is decomposable, and by 1.1, $\tilde{w} = (P^*_{\lambda, X} - \lambda.1)\tilde{w}_1$ for some $\tilde{w}_1 \in [X, K(Z_p, m)]$. Let $[f] = [f'] - j_* w_1$. Then $f: X \to Y$ is also a lifting of f_0 and, as $[\ , K(Z_p, m)]$ acts centrally on $[\ , Y]$,

$$[f \circ P_{\lambda, X}] - [P_{\lambda, Y} \circ f] = [f' \circ P_{\lambda, X}] - j_* P^*_{\lambda, X} \tilde{w}_1 + j_* P^*_{\lambda, K} \tilde{w}_1 - [P_{\lambda, Y} \circ f'] =$$
$$w - j_*(P^*_{\lambda, X} \tilde{w}_1 - \lambda \tilde{w}_1) = 0,$$

and (b) holds for f.

1.3. Corollary. _Let_ X _be an H-space, and_ $H^*(X, Z_p)$ _as in Theorem B. Let_ $y \in QH^{n_k}(X, Z_p)$. _Then there exists a map_ $f: X \to S^{n_k}(p)$ _with_ $y \in \operatorname{im} QH^{n_k}(f)$, _where_ $S^{n_k}(p)$ _is the mod p_ S^{n_k} _(in the sense of_ [6]).

Proof. Let $\tilde{y} \in H^{n_k}(X, Z)$ represent y. $\tilde{y}$ can be chosen so that $(P^*_{\lambda, X} - \lambda.1)\tilde{y}$ is decomposable. By a further alteration of $\tilde{y}$ (e.g. replacing it by $\prod_{1 \leq t < p} (\lambda^t - \lambda)\tilde{y} - \tilde{d}$ for some decomposable $\tilde{d}$) one may assume $P^*_{\lambda, X}\tilde{y} - \lambda\tilde{y} = 0$. Hence, by 1.2, the geometric realization of $\tilde{y} - f_0 : X \to K(Z, n_k)$ can be lifted to $f: X \to S^{n_k}(p)$.

1.4. Corollary. _Let_ P_1 _be a set of primes. Let_ X _be an_

H-space such that, for every $p \in P_1$, $H^*(X, Z_p)$ satisfies the hypothesis of Theorem B. Then for every $y \in Q(H^{n_k}(X, Z)/\text{torsion})$ there exists $\tilde{f} : X \to S^{n_k}$ so that $\lambda y \in \operatorname{im} Q(H^*(f, Z)/\text{torsion})$ with λ prime to P_1.

Proof. $f : X \to S^{n_k}(p)$ of 1.3 satisfies $\lambda y \in Q(H^*(f, Z)/\text{torsion})$ with λ prime to p. Though f does not necessarily lift to $\tilde{f} : X \to S^{n_k}$ but an r multiple of it, i.e. the composition $X \xrightarrow{f} S^{n_k}(p) \xrightarrow{P_r} S^{n_k}(p)$ (where P_r is the r-th power map) does, where r is prime to p. Hence, for every $p \in P_1$ there exists $f_p : X \to S^{n_k}$ with $r_p y \in \operatorname{im} Q(H^*(f_p, Z)/\text{torsion})$ and r_p prime to p. 1.4 follows now from the following.

1.4.1. Lemma. Let $f_{m_i} : X \to S^{n_k}$, $i = 1, 2$ satisfy $QH^*(f_{m_i}, Z)z = m_i y_{n_k}$ where $z \in QH^{n_k}(S^{n_k}, Z)$ is a generator. Then there exists $f : X \to S^{n_k}$ such that $QH^*(f, Z)z = (m_1, m_2)y$.

Proof. Let a, b' be integers satisfying $am_1 + b'm_2 = (m_1, m_2)$. One may assume that b' is even: $b' = 2b$. Let $m : S^{n_k} \times S^{n_k} \to S^{n_k}$ be of type 1, 2, and let f be the composition

$$X \xrightarrow{\Delta} X \times X \xrightarrow{f_{m_1} \times f_{m_2}} S^{n_k} \times S^{n_k} \xrightarrow{h_a \times h_b} S^{n_k} \times S^{n_k} \xrightarrow{m} S^{n_k},$$

($\deg h_a = a$ $\deg h_b = b$).

1.5. Corollary. Let X, Y be H-spaces. Suppose for every $p \in P_1$, $H^*(X, Z_p) \cong H^*(Y, Z_p) = \Lambda(x_{n_1}, \dots x_{n_k})$, $k < \min P_1$. If $X \cong_{P_1} Y$ in $\dim < n_k$, i.e. if there exists a P_1 H-equivalence $f_0 : X_{n_k} \to Y_{n_k}$ (X_{n_k} and Y_{n_k} are the Postnikov approximations), then $X \approx_{P_1} Y$ as spaces.

Proof. Under the hypothesis one can apply theorem B for the map $X \to Y_{n_k}$ to get a lifting $f : X \to Y$ which is an isomorphism of Z_p-cohomology in dimensions less than n_k and therefore in all dimensions.

2. Proof of Theorem A (a) and (b). Let P_1 denote the set of odd primes.

2.1. Proof of Theorem A(a). One may assume that, mod 2, X fibers over S^m, as one may replace X by $X^1 \approx_2 S^n \times S^m$. Hence there exists $f_2 : X \to S^m$ with $H^*(f, Z)z = \lambda y$, where λ is odd. Applying 1.4 and 1.4.1 A(a) follows.

2.2. Proof of A(b). Suppose E_α and E'_α are the total spaces of S^n fibrations over S^m with $S^n \cup_\alpha e^m$ as m-skeleton. Suppose both are mod odd H-spaces. Replace E_α and E'_α by mod p equivalent H-spaces X and Y. We may assume that $\pi(Y)$ has no 2-torsion.

Let $h_0 : Y \to K(Z, n) \times K(Z, m)$ induce an isomorphism of rational cohomology. Decompose h_0 with a Postnikov system $h_{k,k-1} : Y_k \to Y_{k-1}$ with fiber $h_{k,k-1} = K(G_k, k)$, where G_k are odd torsion groups. As $X \approx_{P_1} Y$ in $\dim \leq m + n - 1$ one has a map $f_{m+1} : X \to Y_{m+1}$ inducing an isomorphism of mod p cohomology, $p \in P_1$, in $\dim \leq m + 2$. If we show that f_{m+1} satisfies condition (b_0) of Theorem B then Theorem A(b) will be proved, as f_{m+1} can then be lifted to $f : X \to Y$, which in turn will be a P_1 equivalence (as in 1.5). Hence, we have to prove:

2.2.1. $f_{m+1} \circ P_{\lambda, X} \sim P_{\lambda, Y_{m+1}} \circ f_{m+1}$.

For every positive integer k let $w_{k,s} : X \wedge X \to Y_s$ measure the difference between $\mu_{Y_s}(f_s \times f_s)(1 \times P_{k,X})$ and $f_s\mu_X(1 \times P_{k,X})$. As $f_s : X \to Y_s$ is an H-map for $s < 2n$, $w_{k,s} \sim *$ in this range. Hence $w_{k,s}$, for $s \geq 2n$, lifts to $\tilde{w}_{k,s} : X \wedge X \to Y_{s,2n}$, where $Y_{s,2n}$ is the fiber of $Y_s \to Y_{2n-1}$, and we may assume $h_{s,s-1}\tilde{w}_{k,s} = \tilde{w}_{k,s-1}$. Now $Y_{2n,2n} = K(G_{2n}, 2n)$, hence $[\tilde{w}_{k,2n}] \in H^{2n}(X \wedge X, G_{2n})$; and if $T : X \wedge X \to X \wedge X$ is the twisting (i. e. $T(a, b) = (b, a)$) then $(1 + T^*)[\tilde{w}_{k,2n}] = 0$. As $H^s(X \wedge X, G) = 0$ for $2n < s < n + m$ we have $(1 + T^*)[\tilde{w}_{k,s}] = 0$ for $s < n + m$. In particular $(1 + T^*)[\tilde{w}_{k,m+1}] = 0$, and consequently $0 = \Delta^*(1 + T^*)[\tilde{w}_{k,m+1}) = 2\Delta^*[\tilde{w}_{k,m+1}]$. As $Y_{m+1,2n}$ is an odd torsion space $\Delta^*[\tilde{w}_{k,m+1}] = 0$,

$w_{k,m+1} \circ \Delta \sim *$, and $\mu_{Y_{m+1}}(f_{m+1} \times f_{m+1})(1 \times P_{k,X}) \circ \Delta \sim$ $f_{m+1}\mu_X(1 \times P_{k,X}) \circ \Delta$. As $\mu_X(1 \times P_{k,X}) \circ \Delta = P_{k+1,X}$, if we assume inductively (on k) that

$$f_{m+1}P_{k-1,X} \sim P_{k-1,Y_{m+1}} \circ f_{m+1},$$

then

$$P_{k,Y_{m+1}} \circ f_{m+1} = \mu_{Y_{m+1}}(f_{m+1} \times P_{k-1,Y_{m+1}} \circ f_{m+1}) \circ \Delta \sim$$

$$\mu_{Y_{m+1}}(f_{m+1} \times f_{m+1}P_{k-1,X})\Delta \sim f_{m+1} \circ \mu_X(1 \times P_{k-1,X}) \circ \Delta =$$

$$f_{m+1}P_{k,X},$$

and 2.2.1 is proved.

3. **Homotopy associativity obstructions.** In this section we review some of the ideas in [3] as they were applied in [5] and derive some further homotopy associativity obstructions.

Let X, μ be an H-space. The obstruction for X, μ to be homotopy associative is given by the class α_X in $[X \wedge X \wedge X, X]$ induced by $[\mu(\mu \times 1)] - [\mu(1 \times \mu)]$. If $\alpha = 0$ we say that X is a homotopy associative H space or simply an A-space (A_3 in [3]).

If X, μ and $\tilde{X}, \tilde{\mu}$ are A-spaces and $f : X \to \tilde{X}$ is an H-map then the homotopy associativity obstruction α_Y of the fiber Y of f lies in $[Y \wedge Y \wedge Y, Y]$. But as under the projection $[Y \wedge Y \wedge Y, Y] \to [Y \wedge Y \wedge Y, X]$, α_Y vanishes, α_Y is the image of a class in $[Y \wedge Y \wedge Y, \Omega\tilde{X}]$, or even of a class in $a_f \in [X \wedge X \wedge X, \Omega\tilde{X}]$ (that can be explicitly described in terms of the homotopy associative structures of X and $\tilde{X}$ and the H-structure of f). In case $\tilde{X} = K(G, m+1)$, $a_f \in H^m(X \wedge X \wedge X, G)$. If one alters the H-structure of f by a map $w : X \wedge X \to \Omega\tilde{X}$, a_f is altered by

$$[(\mu \times 1)^* - (p_2 \times 1)^* - (p_1 \times 1)^* - (1 \times \mu)^* + (1 \times p_1)^* + (1 \times p_2)^*][w] = d_1[w].$$

($p_i : X \times X \to X$ are the projections.)

If $\Omega\tilde{X} = K(G, n)$ and $\mathrm{Tor}_Z(H(X), G) = 0$ then

$d_1[w] = (\bar{\mu} \otimes 1 - 1 \otimes \bar{\mu}^*)[w]$.

We shall consider also obstructions $\alpha_{X,\Delta}$ and $a_{f,\Delta}$ derived from α_X and a_f as follows: $\alpha_{X,\Delta}$ will measure the difference between $\mu(\mu(x, y), y)$ and $\mu(x, \mu(y, y))$ for every $(x, y) \in X \times X$. $\alpha_{X,\Delta} = (1 \times \bar{\Delta})\alpha \in [X \wedge X, X]$, where $\bar{\Delta} : X \to X \wedge X$ is induced by the diagonal. If $\alpha_{X,\Delta} = 0$ we shall say that X is an A_Δ space. $a_{f,\Delta}$ is derived from $\alpha_{X,\Delta}$ the way a_f is derived from α_X. Altering the H-structure of f by $w : X \wedge X \to \Omega \tilde{X}$ alters $a_{f,\Delta}$ by $(1 \times \bar{\Delta})^* d_1[w]$.

4. The proof of Theorem A (c). Let $\alpha \in \pi_{m-1}(S^n)$ and put $K_\alpha = S^n \cup_\alpha e^m$. Then

$$S^{n+k} \to \Sigma^k K_\alpha \to S^{n+k}$$

is a fibration in $\dim < 2n + 2k$. Hence, in $\dim < 2n + k$,

$$\Omega^k S^{n+k} \to \Omega^k \Sigma^k K_\alpha \to \Omega^k S^{m+k}$$

is a fibration of k-fold loop spaces. Pulling back by $S^m \to \Omega^k S^{m+k}$ (which is an A-map mod $p > 3$) one obtains a fibration of A spaces, and maps $\Omega^k S^{n+k} \to E_{\alpha,\Omega} \to S^m$, and, for $p > 3$, that $H^*(E_{\alpha,\Omega}, Z_p) = \Lambda(x, y)$ ($\dim x = n$, $\dim y = m$) in $\dim < (n + 1)p - 2$ or in $\dim < 5n + 3$. K_α is the m skeleton of $E_{\alpha,\Omega}$.

To obtain a mod $p > 3$ H-space of type n, m one should construct a sequence

$$E_\alpha \to E_{2n+2m-1} \xrightarrow{g_{n+2m-1}} E_{n+2m-2} \to \cdots \xrightarrow{g_{5n+2}} E_{5n+1} = E_{\Omega,\alpha}$$

of H-spaces and maps, where g_k is a principal $K(G_k, k)$ fibration (G_k is a $p > 3$ torsion group) induced by an H-map $h_k : E_{k-1} \to K(G_k, k+1)$ such that $\operatorname{im} QH^{k+1}(h_k, Z_p) = QH^{k+1}(E_{k-1}, Z_p)$ for all $p > 3$. (Note that $5n + 1 > m$.)

Now for dimensional reasons h_k is an H-map for $2n + m + 2 \leq 5n + 2 < k < 2m + n$. Moreover, for the same reasons, if $2n + m + 2 \leq k \leq 2m + n$, $a_{h_k} = 0$. Hence E_{n+2m-2} is an A-space, and a generator z in $H^{n+2m}(E_{n+2m-2}, G_{n+2m-1})$ will have as coproduct

$$\mu^*_{E_{n+2m-2}} z = z \otimes 1 + 1 \otimes z + j_1 xy \otimes y + j_2 y \otimes xy,$$

where $j_i \in G_{n+2m-1}$. But $(\mu^*_E \otimes 1)\mu^*_E = (1 \otimes \mu^*_E)\mu^*_E$ implies $j_1 = j_2 = 0$ and z is primitive. Hence h_{n+2m-1} is an H-map and E_{n+2m-1} is an H-space. Moreover for dimensional reasons E_{n+2m-1} is an A-space and h_{n+2m} is an H-map. However

$$a_{h_{n+2m}} \in H^{n+2m}(E_{n+2m-1} \wedge E_{n+2m-1} \wedge E_{n+2m-1}, G_{n+2m})$$

does not necessarily vanish.

$$a_{h_{n+2m}} = j_1 x \otimes y \otimes y + j_2 y \otimes x \otimes y + j_3 y \otimes y \otimes x.$$

Altering the H-structure of h_{n+2m}, however, may force j_2 and j_3 to vanish and hence one may assume $a_{h_{n+2m}} = j_1 x \otimes y \otimes y$, with $j_1 \in G_{n+2m}$. Hence $(1 \otimes \overline{\Delta}^*)a_{h_{n+2m}} = 0$, and E_{n+2m} is an A_Δ space. Again for dimensional reasonal reasons E_k is an H-space for $k < 2n + 2m - 2$. Moreover $a_{h_{k,\Delta}} \in H^k(E_{k-1} \wedge E_{k-1}, G_k)$ vanishes in this range as well. Hence E_k is an A_Δ space. In particular $E_{2n+2m-2}$ is an A_Δ space. If $z \in QH^{2n+2m}(E_{2n+2m-2}, G_{2n+2m-1})$ then $\overline{\mu}^* z = jxy \otimes xy$. But because $E_{2n+2m-2}$ is an A_Δ-space, $(1 \otimes \overline{\Delta}^*)(\overline{\mu}^*_{E_{2n+2m-2}} \otimes 1 - 1 \otimes \overline{\mu}^*_{E_{2n+2m-2}}) = 0$, and hence $j = 0$, z is primitive and $E_{2n+2m-1}$ is an H-space. $E_\alpha \to E_{2n+2m-1}$ has a $2n + 2m - 1$ connected fiber, hence the k-invariants are in $\dim > 2n + 2m$ and are all primitive, and E_α is an H-space.

REFERENCES

1. P. Hilton, and J. Roitberg. On the classiciation problem of H-spaces of rank two. Comment. Math. Helvetici 45 (1970), 506-16.
2. M. Mimura, G. Nishida and H. Toda. On classification of H-spaces of rank 2 (mimeographed).

3. J. Stasheff. Homotopy associativity of H-spaces, I, II. Trans. Amer. Math. Soc. 108 (1963), 275-312.
4. J. Stasheff. Sphere bundles over spheres as H-spaces mod $p > 2$. Symposium on Algebraic topology, Battelle, Seattle 1971, Springer Lecture notes 249.
5. A. Zabrodsky. Implications in the cohomology of H-spaces. Ill. J. of Math. 14 (1970), 363-75.
6. A. Zabrodsky. Homotopy associativity and finite CW complexes. Topology, 9 (1970), 121-8.
7. A. Zabrodsky. The classification of H-spaces with three cells I and II. Math. Scan. (to appear).

Hebrew University
Jerusalem
Israel